BLOOD

Blood is life, its complex composition is finely attuned to our vital needs and functions. Blood can also signify death, while 'bloody' is a curse. Arising from the 2021 Darwin College Lectures, this volume invites leading thinkers on the subject to explore the many meanings of blood across a diverse range of disciplines. Through the eyes of artist Marc Quinn, the paradoxical nature of blood plays with the notion of self. Through those of geneticist Walter Bodmer, it becomes a scientific reality: bloodlines and diaspora capture our notions of community. The transfer of blood between bodies, as Rose George relates, can save lives, or as we learn from Claire Roddie can cure cancer. Tim Pedley and Stuart Egginton explore the extraordinary complexity of blood as a critical biological fluid. Sarah Read examines the intimate connection between blood and womanhood, as Carol Senf does in her consideration of Bram Stoker's novel *Dracula*.

IOSIFINA FOSKOLOU is a cancer biologist with joint appointments in Amsterdam and Cambridge. She has authored numerous scientific journal articles and received research awards/grants in Oxford, Cambridge, and the Karolinska Institute, Stockholm. She is a strong advocate of women in science.

MARTIN JONES is an archaeologist with an interest in food and diet. He is editor/co-editor of eight volumes, and the author of 150 scientific papers and four books. Among those books, *Feast: Why Humans Share Food* received the 2008 *Food Book of the Year Award* from the Guild of Food Writers.

THE DARWIN COLLEGE LECTURES

These essays are developed from the 2021 Darwin College Lecture Series. Now in their thirty-seventh year, these popular Cambridge talks take a single theme each year. Internationally distinguished scholars, skilled as popularisers, address the theme from the point of view of eight different arts and sciences disciplines.

Subjects covered in the series include

Blood

Edited by *Iosifina Foskolou*

University of Cambridge

Martin Jones

University of Cambridge

CAMBRIDGE
UNIVERSITY PRESS

CAMBRIDGE
UNIVERSITY PRESS

University Printing House, Cambridge CB2 8BS, United Kingdom

One Liberty Plaza, 20th Floor, New York, NY 10006, USA

477 Williamstown Road, Port Melbourne, VIC 3207, Australia

314–321, 3rd Floor, Plot 3, Splendor Forum, Jasola District Centre, New Delhi – 110025, India

103 Penang Road, #05-06/07, Visioncrest Commercial, Singapore 238467

Cambridge University Press is part of the University of Cambridge.
It furthers the University's mission by disseminating knowledge in the pursuit of
education, learning, and research at the highest international levels of excellence.

www.cambridge.org
Information on this title: www.cambridge.org/9781009205498
DOI: 10.1017/9781009205528

First published 2022

Printed in the United Kingdom by TJ Books Limited, Padstow Cornwall

A catalogue record for this publication is available from the British Library.

ISBN 978-1-009-20549-8 Paperback

Contents

Contents

Figures

Notes on Contributors

Walter Bodmer is a human geneticist and cancer researcher. Formerly professor of genetics at Stanford and Oxford Universities and then Director of the Imperial Cancer Research Fund (now CRUK), he now leads the Cancer and Immunogenetics Laboratory at the Weatherall Institute of Molecular Medicine at Oxford University, where he is an Emeritus Professor. His major current research interests are in the fundamental genetics and biology of colorectal cancer and their potential clinical applications, and in the characterisation, analysis, and population distribution of genetic diversity in human populations.

Stuart Egginton is a cardiovascular and muscle physiologist at the University of Leeds, where he is Professor of Exercise Science. His work explores biological limits to activity, and how flexibility is essential to cope with physiological challenges. He is a Fellow of the Physiological Society, currently a Monitoring Editor for the *Journal of Experimental Biology*, and has served as President of the British Microcirculation Society.

Rose George is a journalist and author, who has tackled a wide range of pressing global issues including refugees, shipping, and human waste. George's career has included senior editorships at *Colors* and *Tank* magazine, and she was a war correspondent in Kosovo for *Condé Nast Traveler* magazine. Her books include *The Big Necessity* (2008), *Ninety Percent of Everything* (2013), and *Nine Pints* (2018).

Tim Pedley is an applied mathematician whose research has been devoted to Biological Fluid Dynamics, both internal (e.g., blood flow) and external (e.g., micro-organism swimming), for over 50 years. He is Emeritus G. I. Taylor Professor of Fluid Mechanics at Cambridge, and has served as Chairman of the World Council for Biomechanics, President of the International Union of Theoretical and Applied Mechanics, and President of the Institute of Mathematics and Its Applications.

Marc Quinn is a leading contemporary British artist. He first came to prominence in the early 1990s, when he and several peers redefined what it was to make and experience contemporary art. Quinn makes art about what it is to be a person living in the world – whether it concerns people's relationship with nature and how that is mediated by human desire; or what identity and beauty mean and why people are compelled to transform theirs; or representing current, social history in his work. His work also connects frequently and meaningfully with art history, from modern masters back to antiquity.

Sara Read is a Senior Lecturer of literary history at Loughborough University. Her expertise is in the cultural and literary representations of the reproductive female body in early modern England. Read co-edits the history of medicine blog *earlymodernmedicine.com*. She is the author of several books, most recently *The Gossips' Choice*, her debut novel, which is founded in her research specialisms.

Claire Roddie is a Consultant Haematologist and Honorary Senior Lecturer in Haematology at University College London (UCL) with a particular interest in chimeric antigen receptor (CAR)-T cells for cancer treatment. Roddie's current role involves pre-clinical development of novel CAR projects, CAR-T cell manufacture and clinical trial design for academic studies at UCL. She is also responsible for the development of a clinical service at UCL hospital to support patients recruited to immunotherapy studies and those receiving CAR-T cells on the National Health Service (UK).

Carol Senf is a Professor at Georgia Institute of Technology who specialises in Gothic Studies. She has written on Stoker, Dracula, Stephen King, Sheridan LeFanu, Mary Shelley, the Brontës, Dickens, Eliot, Hardy, and Sarah Grand. Her most recent book (co-authored with Sherry Brown and Ellen Stockstill) is *A Research Guide to Gothic Literature in English* (2018).

Acknowledgements

These series always take a great deal of organisation, this year with the added challenge of entering the novel and untested landscape of Covid isolation. Much work went into organising the talks in an online format, and we are immensely grateful to the speakers and to our IT team, Espen Koht and Jamie Pilmer, who had to quickly assemble an entirely new way of doing things. The support and guidance of the College Master of Darwin, Mike Rands, and the Registrar, Janet Gibson, and of the Education and Research Committee were critical throughout. Ivan Higney and the College catering team came up with imaginative ways to compensate for the absence of the normal dinner and attendant hospitality. For invaluable support in the assembly of lectures and texts, we should also like to thank Catrin Owen and Cameron Brick. We also warmly thank those donors whose generous financial support has enabled the Darwin College Lecture Series to flourish over the years.

Introduction

IOSIFINA FOSKOLOU AND MARTIN JONES

There are many reasons why we proposed the theme of 'Blood' for the 2021 Darwin College Lecture Series. The first reason was quite personal: blood was the connection between one of us (IF) and Darwin. Darwin College funded – through the Evelyn Trust – Iosifina's research on a novel cancer therapy, which uses blood cells to target cancer. Claire Roddie explains in detail how this transformative therapy works in her chapter. From there we moved to explore how blood could connect a plethora of fields and subjects. Although academic merit was one of the most important factors in identifying speakers, we also wanted to give voice to controversial and unfamiliar topics. For example, although it would have been exciting to have a historian talking about blood in wars, battles, or kingdoms of Europe, we instead chose Sara Read, who talked about menstrual blood in early modern England. It is remarkable how little this subject is discussed and how little we know about menstrual health – even in our times – given that half of the Earth's population experiences this monthly cycle. Similarly, when we were thinking about literature, while Bram Stoker's *Dracula* is a predictable choice, we decided to invite Carol Senf, who was one of the first academics to explore feminist themes within that famous text.

Other issues to which we wanted to give voice in this series were social justice/social awareness and climate crisis. Rose George and Marc Quinn seemed the ideal speakers to address social justice and social awareness. Rose George gave voice to heroes and villains of blood. On the one hand, she exposed cruel experiments people used to do in animals until they understood the science behind transfusions, and on the other hand, she reminded us of such heroes as Dame Janet Maria Vaughan, the founder of the UK's blood donation service (NHS Blood and Transplant) as we know it today. Marc Quinn was chosen not only because of his celebrated work *Self* (shown on the cover of this book) but also for his upcoming exhibition 'Our Blood'. This major public artwork aims to raise awareness of the global refugee crisis. Quinn's work touches upon such social issues as disabilities, social exclusion, and racism. In relation to climate crisis, we couldn't have chosen a better speaker than Stuart Egginton. In his many trips to Antarctica Egginton has confronted at first hand the challenges the Antarctic ecosystem faces due to global warming.

Lastly, we wanted to emphasise the ability of blood to connect us. We chose Walter Bodmer and Tim Pedley for this area. Bodmer was one of the first scientists to suggest the Human Genome Project, subsequently leading the People of the British Isles project. From a different scientific angle, Tim Pedley discussed the ability of blood to connect the different tissues within our body and the physical requirements for making it possible for this fluid to keep us alive.

The organisation of the series began with much excitement. Our speakers quickly accepted our invitations to lecture, and everything was in place by the time the Darwin College Lecture Series of 2020 had reached completion. However, this was also the time of the outbreak of the COVID-19 pandemic. We had no idea how long the pandemic would last. As summer approached, it became increasingly clear that our 2021 series would not be able to take place in its normal format. Instead, the entire lecture series was delivered virtually, with prompt and brilliant support from staff at Darwin College and Cambridge University. That happened, and all of the lectures were successfully delivered. One of our speakers, whose main work was at the frontline of Covid mitigation, had to break off mid-lecture to attend to urgent business. With skilful editing from our Darwin team, our audience remained unaware. Our audience remained loyal, and at the time of writing, the series has received over 14,000 views.

Having outlined above themes in the organisers' minds when assembling the series, we can turn to what we have learnt from the resulting lectures. It is one of the stimulating features of the Darwin College Lecture Series that, although each year it is bound by a single word, additional new themes emerge from what is actually assembled and delivered. A theme that proceeds most directly from the conception of the 2021 series and IF's own research is that of blood's physiological function, in ourselves and in other animals. The chapters of Pedley, Egginton, and Roddie bring to light the sheer complexity of this most familiar of fluids. Blood is a complex heterogeneous fluid that is routinely required to follow a complicated path around an animal's body, somehow contained within a circulatory system while effecting continuous physical and chemical exchange across that system's boundary walls. Not only is the sheer mechanical task, as explored in Pedley's chapter, considerable; so is the thermoregulatory task, examined in Egginton's chapter. Managing and sustaining this complex attribute of life, especially in our own species, is at the heart of much medical science, a theme developed by Roddie in the context of her work on leukaemia.

Roddie's chapter draws our attention to the prominent role in medicine played by the transfer of blood from one individual to another, a theme enlarged upon and given historical context in George's account of the varied actors in the history of blood transfusion. George expands from a historical treatment of the transfer of blood to probe at the fluid and diffuse boundaries of her theme. She describes how, since the ancient beginnings of medicine, blood has been perceived as a fundamentally mysterious entity, that can either kill us or save us, ideas that turn our attention from medicine to mythology, from the facts of physiological science to the narratives of folklore and fiction. The latter theme of fiction is further developed by Senf in one of the best-known fables about the movement of blood between bodies.

Senf's reading of Bram Stoker's *Dracula* is not as far removed from 'science' as a subsequent corpus of derivative B-films might lead us to imagine. Indeed, the contrast in Stoker's novel between the blood of the aristocratic past and that of the scientific future seems to allude to the future potential of medical blood transfusion; some of his language suggests Stoker was aware of early attempts at developing an effective procedure. Senf explores the rich variety of meanings that may be attached to the transfer of blood from one body to another, and places *Dracula* at the paradoxical cusp between looking back and looking forward. Prominent within the past are superstition and aristocratic men. Prominent in the future is the New Woman, freeing herself from the shackles of patriarchal Victorian life. Senf is particularly interested in how Stoker's blood-sucking novel of 1897 looks back upon a disappearing century of stifling gender relations and tentatively forward to a new century ahead, and new possibilities of womanhood.

In various of the volume's chapters, the theme of blood in relation to womanhood recurs.

We learn how intimately blood is entangled in female identity, physiology, and development, and in different cultural perceptions of each of these, and of the powers of good and evil ascribed to them. As Read explains, culturally specific interpretations of the patterns of bleeding at different stages of a woman's life are repeatedly employed to define those stages, and the attendant rites of passage that punctuate the female biography. At the centre of this discourse is menstrual blood, which, as George explains, has a rich cultural history in its own right. The Biblical Isaiah and Pliny the Younger seem to share a horror of the fluid, the latter offering explanations of how a menstruating woman can kill a swarm of bees, and turn iron and brass rusty. In case we imagine such attitudes are lost in the mists of time, George reminds us that, little more than a decade ago, a (male) British Prime Minister seemed unable to utter the word 'tampon' in Parliament, and as recently as 2019, the world's multi-million-dollar advertising industry had great difficulty coming to terms with menstrual blood's true horrific colour.

The associations between blood and womanhood can be viewed through a wide variety of lenses, ranging from the science to imagination and fear. Through the same range of lenses, blood may be viewed in relation to other dimensions of identity, in particular to ancestry, nationhood, and belonging. In the middle ages, that association might also be regarded as 'scientific'. In thirteenth-century England, one meaning of 'blood' or 'blod' was 'person of one's family, race, kindred, offspring', a meaning shared by Latin *sanguis* and Greek *haima*.[1] That sense is also conserved within such modern usages as 'blue-blooded' and 'blood line'. Blood as ancestry is one of the key themes Senf discerns in Dracula, and the words 'blood line' feature in the title of Bodmer's contribution to this volume.

Bodmer discusses and explains modern techniques of genetics that have long displaced mediaeval notions of blood inheritance. Nonetheless, blood

---

[1] www.etymonline.com/word/blood.

retains a number of roles in his account, most notably in his discussion of pioneering work on blood group genetics, which lies at the foundation of our contemporary understanding of human diversity across the world. This diversity may be resolved in a rich world prehistory that itself can be inferred from that same body of genetic evidence.

It may seem paradoxical that a mediaeval notion of blood ancestry can feature in a cutting-edge account of the genetic science that replaced it, but paradox in itself is a theme to which several authors on the topic of blood return. The manner in which Stoker's text of *Dracula* plays with the paradoxical nature of blood is a recurrent theme of Senf's contribution. 'Blood' can denote the stuff of noble life, while 'bloody' can serve as a curse. The blood of Dracula can allude to strange masculine entities from the past, but also be entangled with the 'New Woman' of a more liberated and egalitarian future. George's chapter also touches on paradox, reminding us that, while blood can save us, it can also kill us. The theme of paradox is especially evident in the conversation with Marc Quinn.

Quinn reflects on the life–death paradox; blood is the essence of life, and it remakes itself, but usually associated with death, violence, and illness. Quinn explores how this theme of paradox moreover applies to the notion of 'self', a notion that has little meaning without acknowledgement of connections that reflect profound interdependence. That collective existence may be captured in the notion of 'blood line', until we acknowledge that the notion of blood line is framed as much around exclusion as around inclusion, returning us once again to the theme of paradox.

So, our exploration of 'blood' emerged by combining a medical research interest in a tangible, vital fluid with concerns of social justice and climate change. Our lecturers picked up those themes and went on to explore themes ranging from blood lines, identity, and womanhood, to the theme of enduring paradox in relation to past and future, to self and community. Such is the intrigue of the Darwin College Lecture Series' tradition of exploring a single word and seeing where it takes us.

1 Battle Blood

CLAIRE RODDIE

The subject of 'Battle Blood' is the treatment of the blood cancer leukaemia. I am a haematologist by training, and treating blood cancer is my daily job. In the present day we are fortunate to have sophisticated methods to diagnose and treat blood cancers, but this was not always the case. The celebrated poet Hilaire Belloc (1907) reflects the experience of haematologists at the beginning of the last century, when they were faced with the prospect of treating leukaemia:

> Physicians of the Utmost Fame
> Were called at once; but when they came
> They answered as they took their Fees,
> There is no Cure for this Disease.

In fact, physicians did not really understand leukaemia, certainly not in 1845 when John Bennett, a Scottish physician, described an unusual case of a 28-year-old slate layer who presented with gross fatigue and a large tumour in the left side of his abdomen (Bennett 1845). This patient developed fevers, bleeding, and abdominal pain, and multiple new tumours appeared in his armpits, groin, and neck, correlating anatomically with the lymph node regions in the human body. The best therapy for this 'condition' in 1845 was a combination of leeches and purging, but unsurprisingly the patient did not respond to these 'treatments' and died from progressive disease a mere few weeks later.

In an attempt to better understand this unusual syndrome, Bennett performed an autopsy and, when he examined the blood, he found it was full of white blood cells, the principal constituent of pus. The function of white blood cells is to protect the body from infection, so naturally Bennett presumed that this patient had died from an overwhelming sepsis. However, the puzzling aspect was that he could not find a source of infection. He simply concluded that it was a 'suppuration of pus with no clear cause'.

A few years later, Rudolf Virchow, a German researcher, described a 50-year-old lady with an excess of white cells in the blood and a huge spleen, but again no signs of overt infection (Virchow 1856). Virchow concluded that this must be an intrinsic problem with the white blood cells. The word 'leukaemia' is derived from 'leukos', the Greek word for white.

Virchow began to develop systems to classify biological anomalies, including aberrant cell growth. He defined hypertrophy as an increase in cell size, and hyperplasia as an increase in cell number. For Virchow, leukaemia was a pathological, unexplained, abnormal hyperplasia of white blood cells.

In essence, the condition we know today as B-cell acute lymphoblastic leukaemia (B-ALL) is essentially an uncontrolled expansion of immature white blood cells, much as Virchow described it. These immature cells are unable to perform the functions of mature white blood cells, that is, to protect the host from infection. Worse still, the leukaemia cells destroy the normal architecture of the bone marrow and completely disrupt normal blood production. This commonly leads to symptoms such as anaemia, low platelets, and a reduction in normal (functional) white blood cells. Anaemia is the occurrence of low red blood cell levels, and a deficiency in these oxygen-carrying cells can lead to breathlessness and fatigue. The function of platelets is to stop bleeding, and self-evidently, when the platelet count is low, the risk of spontaneous and uncontrollable bleeding is increased. Where the normal white blood cell levels are reduced, severe infection is the likely outcome.

Patients describe myriad other symptoms relating to the rapid growth of leukaemia cells, including fever, weight loss, and bone pains. Indeed, examination of leukaemia patients often reveals enlarged lymph nodes and an enlarged spleen, with associated pallor and bruising.

By the early part of the twentieth century, little progress had been made in the treatment of leukaemia, and the physician's perspective in 1950 wasn't particularly different from the view in the 1840s. William Castle describes acute lymphoblastic leukaemia as follows: 'Its palliation is a daily task, its cure a fervent hope' (Mukherjee 2010).

However, one man helped change the landscape of acute lymphoblastic leukaemia. Sidney Farber worked in Boston as a pathologist at the

Children's' Hospital. He had written a book on the classification of childhood tumours, called *The Post-mortem Examination* (Farber 1937), and developed a scientific interest in what he described as the 'hopeless' condition of childhood leukaemia. At that time there was no national cancer research strategy, and treatment options were limited to radiation and surgery, both of which offer limited value in a blood-borne disease.

Farber reasoned that, to develop better treatments for cancer, research to understand and classify cancer was required. In order to understand cancer, he proposed systems to measure it, and he reasoned that leukaemia was an excellent model system, as white blood cells can easily be measured in the blood. If cancer can be measured, then one can begin to investigate the impact and potency of therapeutic interventions in living patients. If white blood cells grow or die, then that in itself is a measure of the success (or failure) of a proposed therapy.

Farber looked for ideas and inspiration for potential therapies from many sources. He discovered the work of Lucy Wills, an English physician, who in 1928 travelled to Bombay to study the profound anaemia that was observed in factory workers there. Those affected were often malnourished, and, more often than not, the anaemia seemed to specifically affect mothers and their children more than other groups.

By a surprising turn of events, Wills discovered that the anaemia was cured with Marmite (Wills 1933). And when she investigated this, she found that the active constituent of Marmite is folic acid, a vitamin found in fruit and vegetables that is essential for healthy blood production. Indeed, when we grow and repair tissues in our bodies, cells make copies of themselves and, to do this, they must make copies of their DNA. Adequate folic acid is critical for DNA production. Healthy people make more than 300 billion blood cells every day, so it is incredibly important that folic acid levels are maintained within the normal range.

Applying Lucy Will's findings regarding folic acid to acute leukaemia, Sidney Farber wondered whether folic acid supplementation would improve the low blood counts associated with acute leukaemia. He reasoned that the leukaemia cells were consuming all available folic acid, leading to a state of folic acid deficiency in the residual, 'normal' cells, stopping them from growing properly.

To test this hypothesis, Farber instituted an early clinical trial using synthetic versions of folic acid in paediatric patients with acute leukaemia. Contrary to his expectations, he found that folic acid supplementation seemed only to accelerate leukaemia growth. Whilst this was disappointing, it led him to his second research question: if supplementing folic acid was leading to rapid leukaemia progression, could it be that antagonising or blocking of folic acid would have the opposite effect?

Farber had a close collaborator and friend, Yellapragada Subbarow, a biochemist who was also trained as a physician. Subbarow's biochemistry laboratory was focused on the synthesis of synthetic versions of compounds and chemicals that exist within normal cells, including attempts to create synthetic versions of folic acid. During this process he inadvertently generated a range of folic acid antagonists/blocking molecules, including one called aminopterin (Farber et al. 1948; Mukherjee 2010).

Working with Subbarow, Farber proposed a clinical trial of aminopterin, to block folic acid, for childhood leukaemia. The first patient treated with aminopterin was Robert Sandler, a boy with childhood leukaemia complicated by bone disease and fractures. Following aminopterin, Sandler very quickly achieved a dramatic remission, which was unprecedented in the field of acute leukaemia.

Unfortunately, the remission was short-lived, and he relapsed after several months, but the finding that a single, simple chemotherapy drug could so dramatically control acute leukaemia was a paradigm shift in the field and sparked a new wave of research efforts into many different types of chemotherapy for this condition.

The second huge advance in the field of acute leukaemia was the inception of allogeneic bone marrow transplantation. One of Sidney Farber's protégés, Edward Donnall Thomas, won a Nobel Prize in 1990 for conceiving of this therapeutic approach for the treatment of blood cancers and is known as the father of bone marrow transplantation. Donnall Thomas showed that an obliterative dose of chemotherapy/radiotherapy could eradicate the patient's immune system and that the patient's blood production could be 'rescued' with transplanted bone marrow, taken from a healthy donor (Thomas et al. 1957; 1975).

The first successful allogeneic bone marrow transplant was performed in 1956, in a twin boy with acute leukaemia. He received obliterative radiotherapy followed by a healthy bone marrow donation from his identical twin and survived. This early success prompted a huge increase in bone marrow transplant research.

Dr George Mathé, a French oncologist and immunologist, was also fascinated by allogeneic bone marrow transplant. He applied this novel approach in six nuclear reactor engineers with severe and debilitating bone marrow aplasia following a nuclear reactor incident. Since they did not have the option of identical twin donors, Mathé gave them bone marrow derived from an unrelated donor. Unexpectedly, all the patients developed an unexplained, debilitating wasting condition following the donor cells, which we now know as graft versus host disease (GvHD) (Mathé et al. 1959). This life-threatening condition arises where the incoming immune cells in the bone marrow donation are not well matched to the patient's own immune system. These mismatched cells can thus recognise the patient as 'non-self' and attack their normal cells and tissues, leading to a devastating immunological treatment complication.

Dr Jean Dausset, a French immunologist, made a critical discovery, identifying the reason for the GvHD phenomenon, for which he won the Nobel Prize in 1980. He described a mismatch in the protein signature on the cell surface of the blood compartment of the patient compared with the donor, referred to as the human leukocyte antigens (HLA) mismatch (Dausset 1958). This mismatch in HLA between donor and patient is the reason both for the immunological rejection of the donor cells by the patient's immune system and for the immunological 'rejection' of the patient's normal cells/tissues by the donor immune system, i.e. GvHD. He reasoned that HLA compatibility is key to the success of allogeneic bone marrow transplant.

With significant further research and development, by the 1980s, doctors were routinely using allogeneic bone marrow transplant for chemotherapy- and radiotherapy-refractory blood cancers.

In the modern day, standard therapy for acute (lymphoblastic) leukaemia is chemotherapy-based. The first step is induction therapy, where newly diagnosed patients receive multiple chemotherapy drugs as an inpatient in hospital over one to two months. This therapy renders the patients aplastic, with no functioning immune system. To protect these vulnerable patients from infection, we isolate them in single rooms in hospital wards.

The next phase is consolidation, which is a lengthy process for patients who have achieved a good response to induction, followed by maintenance chemotherapy for two to three years. The role of consolidation and maintenance therapy is to try to prevent the leukaemia from coming back. In some cases of high-risk acute leukaemia, we use allogeneic bone marrow transplantation to help prevent relapse.

Despite advances, the battle with acute lymphoblastic leukaemia is far from over. In the paediatric setting (children and young adult medicine), acute lymphoblastic leukaemia is the most common childhood cancer, and for the 20 per cent of patients who relapse following standard therapy, the overall survival rate is approximately 40–50 per cent. In adult patients, relapsed acute lymphoblastic leukaemia is even more challenging. For the 50 per cent of patients who relapse following standard therapies, including allogeneic bone marrow transplant, the five-year overall survival is as low as 7 per cent.

New treatments for leukaemia that has failed to be resolved by bone marrow transplantation and chemotherapy are urgently needed, and this is where redirecting the immune system to recognise and target leukaemia becomes particularly interesting.

T-cells are immune cells that selectively recognise and kill virus-infected cells by binding to them, punching little holes in their surface, and injecting toxins into them. Researchers have long sought to harness this potent cytotoxicity, and to redirect it towards cancer.

To date, T-cell therapies have been used with some success to treat virus-associated cancers, including virus-associated lymphoma. To summarise, virus-reactive T-cells are extracted from patient blood, cultivated in the laboratory, and then administered back to the patient in large cell numbers.

Similarly, high numbers of T-cells can be found within certain cancer tissue biopsies, including tumours such as metastatic melanoma. Tumour-infiltrating lymphocytes (TILs) are thought to have the ability to recognise and kill tumour cells for certain tumour types. However, most cancers cannot be easily recognised by the immune system because cancers arise from 'self' cells, and our immune systems have evolved to tolerate 'self' to avoid the risk of autoimmunity. We believe that the answer to this problem of tumour recognition by the immune system lies in gene engineering and synthetic biology.

Researchers have shown that you can take existing components from within the immune system, including B-cells, which make antibodies to target/bind bacteria with high specificity, and T-cells, which kill virus-infected cells, and, by combining their respective functions, we can couple T-cell killing capability with exquisitely specific targeting of cell surface proteins.

Using molecular cloning, we can insert B-cell genes ('generating' an antibody) into T-cells to create what is known as a chimeric antigen receptor (or CAR for short). The word chimera derives from Greek mythology, and describes a creature that is a fusion of a lion's head, a goat's body, and a serpent's tail. The term CAR reflects the paradoxical expression of a component of a B-cell (antibody) on a T-cell.

A CAR is a very simple structure and easy to design. It comprises three portions: an antibody-derived domain that can target any cell surface protein; a hinge/spacer region to project the antibody region off the surface of the cell; and a portion inside the T-cell that sends signals into the cell to divide and to kill. To make CAR T-cells for patients, you take blood from the patient, extract the T-cells, use a safety-modified viral vector to introduce a new gene into the T-cells, expand them in the laboratory, and then inject them back into the patient.

These super-charged killer CAR T-cells can be used to target cancer cells. A single infusion of this living drug can divide and multiply into millions of cells inside the patient and persist for many years.

So far, much work has been done in targeting a specific protein on B-cells called CD19. CD19 is a good target for CAR T-cells as it is expressed at really high levels on B-cells, but not other cells, so the risks of off-tumour toxicity are low.

However, initial work with CD19 CAR T-cells was fraught with challenges. Original 'first-generation' CAR designs did not expand well in patients, and research showed that early designs had sub-optimal T-cell signalling, so that cells did not proliferate very well when they encountered the tumour (Eshhar et al. 1993). Advanced gene engineering to create second- and third-generation CAR T-cells overcame the limited expansion problems observed with early designs (Friedmann-Morvinski et al. 2005).

Another important factor for CAR T-cell expansion in the patient is the impact of the patient's other immune cells. All cells compete for space, nutrients, and growth chemicals called cytokines to grow well. In order to maximise the space, nutrients, and cytokines available to the incoming CAR T-cells, it was observed that depleting the other (often immunosuppressive) immune compartments using chemotherapy prior to CAR T-cell infusion enhanced their growth and improved their ability to reject tumour cells. We now routinely employ chemotherapy prior to CAR T-cell infusion for this purpose.

A third critical element is how the CAR T-cells are made in the laboratory. Over the last 10 years we have come to understand that the manufacture method and how T-cells are handled in the laboratory is critically important to the clinical outcome. CAR T-cell manufacture is conducted in a sterile 'clean-room' environment. The process is labour-intensive, costly, and biologically complex. The best CAR T-cell products are those that have the capacity to expand in the patient to reject the cancer, but also to lay down CAR T-cell 'memory', forming part of the immune system and preventing future relapse. Much research effort is focused on generating CAR T-cell products with memory populations as we know that these will expand and persist better in our patients, potentially leading to better clinical responses (Fraietta et al. 2018).

CAR T-cell treatment is an intensive procedure, and, following infusion, most patients will stay in hospital so that we can manage the potential early side effects of treatment. CAR T-cells travel around the whole body, and they can thus cause problems with multiple organ systems. One of the major side effects is cytokine release syndrome (CRS), where CAR T-cells secrete inflammatory factors into the blood to try to kill the tumour, which can lead to a protracted high temperature, often associated with a drop in blood pressure, kidney failure, respiratory failure, and multi-organ failure. Thankfully, we have access to medications which can temper this immunotoxic process.

Professor Carl June from the University of Pennsylvania in the USA, who is widely regarded as the father of CAR T-cell therapy, found that CRS was associated with a steep rise in the blood levels of the inflammatory cytokine interleukin-6 (IL-6) (Grupp et al. 2013). He reasoned that blocking IL-6 would ameliorate CRS, and explored the use of Tocilizumab, an IL-6 blocking monoclonal antibody, licensed for juvenile arthritis, for this purpose. Tocilizumab was not licensed for CAR T-cell therapy, but the Institutional Review Board at the University of Pennsylvania gave permission to use this drug for life-threatening CRS. Indeed, the impact of Tocilizumab in CRS was remarkable, with resolution of fever in a matter of hours post-dose.

Another challenging and common complication associated with CAR T-cell therapy is neurotoxicity, or irritation of the brain. It seems to be associated with CAR T-cell expansion rather than with the presence of brain involvement by blood cancer per se. A sensitive way of looking for early neurotoxicity is to ask the patient to write a single sentence, which we routinely do daily on the wards, as we find that the quality (and speed) of handwriting is one of the first signs of neurotoxicity, and this allows us to intervene earlier, in an attempt to prevent more severe syndromes. Patients often experience word-finding difficulties: some patients will repeat the same word multiple times without necessarily realising it, and in some cases patients will be (transiently) unable to speak at all. We have even looked after patients who have had seizures and coma as a result of the CAR T-cells, but this is rare. Like with CRS, we have several drugs to suppress the immune system that we use in this setting, including corticosteroids and other more experimental treatments like anakinra, where we block the interleukin-1 (IL-1) pathway.

Longer-term side effects include B-cell aplasia. This predictable side effect describes the phenomenon of the CAR T-cells killing normal B-cell populations in patients, as the CAR cannot distinguish a cancerous B-cell from a normal B-cell, and indiscriminately kills both. As the CAR T-cell is a living drug, it will continue to delete healthy B-cells as long as it persists in the body. The consequence of B-cell aplasia is the inability to make antibodies which protect us from infection and allow us to form immune responses to vaccines. For this reason, CAR T-cells can leave us immunosuppressed for as long as they persist within our bodies. In specific situations we will start regular antibody infusions to try to replace antibody levels in our patients.

Despite the toxicity, the outcomes following CAR T-cell therapy have been truly transformative in the blood cancer space. The poster-girl for CAR therapy is Emily Whitehead, a young girl treated by the University of Pennsylvania. She had chemotherapy- and radiotherapy-refractory leukaemia; her disease was multiply relapsed, and she effectively had advanced, incurable leukaemia at the point of referral for CAR-T. In April 2012 she received just one CAR T-cell infusion and by day 3 post-infusion she was admitted to the Intensive Care Unit with renal, lung, and cardiac failure and a very high fever with no obvious infectious cause. When the team screened her blood for cytokines, they found an elevation in IL-6, 1,000 times the normal levels, and this prompted Carl June to use Tocilizumab. She subsequently defervesced, making a full recovery, and has achieved a long-term remission from her refractory leukaemia (Grupp et al. 2013).

In a larger cohort of paediatric patients, similar initial responses were observed, and, at 12 months, around 50 per cent of patients infused were still in remission, which is unprecedented for any other therapies in this refractory patient cohort. Similar excellent results have been reported using CAR T-cells in patients with aggressive B-cell lymphomas (Chavez et al. 2019). And I think we're all aware of the publicity that this therapy has generated across the world. On the strength of these data, the FDA and EMA have licenced the first CAR T-cell products for use in the USA and on the National Health Service.

In assessing response to CAR T-cell treatment, we often check the CAR T-cell levels in the blood to understand whether they have adequately expanded and can persist. Like Sidney Farber, we find that measuring the blood helps us to determine the success/failure of therapy.

Indeed, patients can relapse with CD19-positive disease, which indicates that the CAR has failed to persist in the blood, possibly rejected by the host immune system, or has become 'exhausted', a process commonly driven by the tumour itself or by poor-quality T-cells from heavily pre-treated patients. There is a huge research effort to overcome these issues, including engineering approaches to make the CAR structure less easy for the host immune system to reject and to manufacture CAR towards better 'T-cell fitness' in an attempt to prevent T-cell exhaustion.

CD19-negative relapse accounts for the majority of cases of relapse. Through some sort of selection pressure exerted by the CAR T-cells, the leukaemia cell loses the target. The CAR continues to circulate in the blood and to kill normal B-cells, but the leukaemia has become invisible, and inevitably repopulates the entire bone marrow with disease. Attempts to overcome antigen escape include targeting several proteins simultaneously, but data are not available yet to confirm whether dual antigen targeting can overcome antigen loss mechanisms.

To conclude, CAR T-cells represent the biggest paradigm shift in malignant haematology in a decade and are now a major therapy for relapsed and refractory leukaemia and lymphoma. They are not without toxic side effects, but we can manage these very well with modern approaches. Most importantly, a significant portion of the patients that we treat can achieve long-term responses following a single dose of CAR T-cells.

The field is rapidly developing, and we are moving further into and beyond the CD19 space, targeting other haematological malignancies and solid cancers. It is just the beginning of the journey.

Of course, the CAR story really is one full of setbacks and roadblocks. If we revisit the experience of researchers in 2008–2009 using first-generation CARs, it's thanks to their resilience and perseverance in going back to the drawing board and back to the lab to move things forward that we are able to offer these transformative medicines to our patients today. We look forward to a future where immunotherapies can replace conventional toxic chemotherapies, not just for leukaemia/lymphoma, but for all cancers.

References

Belloc, H. (1907) 'Henry King'. In *Cautionary Tales for Children*. London: Duckworth.

Bennett, J. H. (1845) 'Case 2, case of hypertrophy of the spleen and liver, in which death took place from suppuration of the blood'. *Edinburgh Medical and Surgical Journal* 64, 413–423.

Chavez, J. C., Bachmeier, C., and Kharfan-Dabaja, M. A. (2019) 'CAR T-cell therapy for B-cell lymphomas: Clinical trial results of available products'. *Theoretical Advances in Hematology* 10, 2040620719841581.

Dausset, J. (1958) 'Iso-leuco-anticorps'. *Acta Haematology* 20, 156–166.

Eshhar, Z., Waks, T., Gross, G., and Schindler, D. G. (1993) 'Specific activation and targeting of cytotoxic lymphocytes through chimeric single chains consisting of antibody-binding domains and the γ or ζ subunits of the immunoglobulin and T-cell receptors'. *Proceedings of the National Academy of Sciences* 90, 720–724

Farber, S. (1937) *The Post-mortem Examination*. Springfield, IL: C. C.Thomas.

Farber, S., Diamond, L. K., Mercer, R. D., Sylvester, R. F. Jr., and Wolff, J. A. (1948) 'Temporary remissions in acute leukaemia in children produced by folic acid antagonist, 4-aminopteroyl-glutamic acid (aminopterin)'. *New England Journal of Medicine* 238, 787–793.

Fraietta, J. A., Lacey, S. F., Orlando, E. J., Pruteanu-Malinici, I., Gohil, M. et al. (2018) 'Determinants of response and resistance to CD19 chimeric antigen receptor (CAR) T cell therapy of chronic lymphocytic leukemia'. *Nature Medicine* 24(5), 563–571.

Friedmann-Morvinski, D., Bendavid, A., Waks, T., Schindler, D., and Eshhar, Z. (2005) 'Redirected primary T cells harboring a chimeric receptor require costimulation for their antigen-specific activation'. *Blood* 105, 3087–3093.

Grupp, S. A., Kalos, M., Barrett, D. Aplenc, R., Porter, D. L. et al. (2013) 'Chimeric antigen receptor-modified T cells for acute lymphoid leukemia'. *New England Journal of Medicine* 368, 1509–1518.

Mathé, G., Jammet, H., Pendic, B., Schwarzenberg, L., Duplan, J. F. et al. (1959) 'Transfusions et greffes de moelle osseuse homologue chez des humains irradiés à haute dose accidentellement'. *Revue Française d'Études Cliniques et Biologiques* 4(3), 226–238.

Mukherjee, S. (2010) *The Emperor of All Maladies*. New York: Scribner.

Thomas, E. D., Lochte, H. L., Lu, W. C., and Ferrebee, J. W. (1957) 'Intravenous infusion of bone marrow in patients receiving radiation and chemotherapy'. *The New England Journal of Medicine* 257, 491–496.

Thomas, E. D., Storb, R., Clift, R. A., Fefer, A., Johnson, L. et al. (1975) 'Bone-marrow transplantation (second of two parts)'. *New England Journal of Medicine* 292, 895–902.

Virchow, R. L. K. (1856) 'Die Leukämie'. In *Gesammelte Abhandlungen zur wissenschaftlichen Medizin*. Frankfurt: Meidinger, pp. 190–212.

Wills, L. (1933) 'The nature of the haemopoietic factor in Marmite'. *Lancet* 221, 1283–1285.

2 Transitional Bleeding in Early Modern England

SARA READ

This essay is drawn from the lecture I gave as part of the 2021 Darwin Lecture Series, which had the theme of 'blood', and will outline the most commonly promoted medical theories as to why women experienced a monthly bleed, present the many words and circumlocutions early moderns used to describe menstruation and related female reproductive bleeding, and discuss prevailing cultural expectations about this event.

In Tudor and Stuart England each new episode of vaginal bleeding was imbued with meaning which related not just to the physiological changes it figured in the female body, but to cultural and social expectations too. This was because each type of bleeding, from menarche to postpartum bleeding, marked a change in the way that a woman was perceived both by herself and by those around her. This paradigm was passed down from ancient medicine, and the Hippocratic text *On Generation*, as Helen King has explained, indicated that, as a girl grows, the channels in her body are gradually opened to make 'a way through and a way outside'. Therefore, as part of her growth to maturity, all 'three transitional bleedings – menarche, defloration and childbirth – cause further changes in the body' (King 1998, 71). In early modern England, the humoral bodily economy as set out in the Hippocratic texts, and modified by Galen in the first century CE, still formed the basis of cultural and medical understandings of the body; by studying early modern medical discourse alongside other more traditionally literary works, we can see that the model of each discreet occasion of female reproductive bleeding marking a change in a woman's body held in the early modern era too. The process of growing from a girl into a woman was still considered to be a gradual one as her body went through a series of physiological changes, but the onset of menstruation marked a girl's transition to young womanhood, and

18

postpartum bleeding signified a change to motherhood. In terms of the reasons why the female body had these episodes of bleeding at all, medical debates were heated, with at least three competing theories claiming to explain it.

The physiology of menstruation might be timeless, but the experience of female reproductive bleeding (from menarche to menopause) is mediated through different cultural norms at any given time. So, when people asked why I was researching this topic, 'Was it different then?,' the answer was yes. It is not different in terms of physiology, but in the way that physiology was understood and experienced by a different society in a different era. Roy Porter has explained that '[t]o a large degree our sense of our bodies, and what happens in and to them, is not first-hand but mediated through maps and expectations derived from the culture at large' (Porter 2003, 44–45), and this is a two-way process, whereby cultural norms inform understandings of the ways a person assumes their body to function, and understandings are reflected into cultural literary and other outputs. The early modern era was one which predominantly saw the body as a humoral one in which blood was one of the main four humours, and men and women's bodies naturally had a different humoral constitution.

One of the main differences in assumptions was that, unlike modern medical assumptions, the Hippocratic taxonomy, as understood in the early modern period, considered that all aspects of vaginal bleeding, including the blood lost upon first intercourse and that lost after childbirth, were directly related to menstruation, and menstrual blood is what is lost on these occasions.

There existed very many euphemisms, circumlocutions, and often just alternative names for menstruation in the Tudor and Stuart eras. The most common, and oldest, of these was 'flowers'. It comes from a horticultural metaphor that without flowers there can be no fruit (Crawford 1981, 51). The next most common names were 'terms' or 'courses', and these names emphasised the periodic nature of the flow. For most of the seventeenth century, medics preferred to use *menses*, from the Latin for monthlies, or used the term *purgations*. From the late seventeenth century, in line with a general move away from Latin amongst scholars, the medical corpus changes to terms based on Greek instead, so we start to see menstruation referred to as *catamenia* (Read 2013, 28). The most normal, formal term nowadays – menstruation – first appeared in print around this time too. In 1686 Gideon Harvey, in *The Conclave of Physicians*, presented an argument between doctors about whether a woman with menstrual problems should be prescribed a cordial or not, and said that he was referring to a case of menstruation (Read 2013, 32). Sometimes women used code names that were personal to them. For example, Queen Anne, who exchanged correspondence with Sarah Churchill, Duchess of Marlborough, referred to her menstrual periods as Lady Charlotte's visits. And other women used even more vague euphemisms like 'those'. Other significant terms or phrases which we use today had no currency in the early modern era. Menarche, for example, is recorded in the *Oxford English Dictionary* only from 1900. Instead, this event was simply referred to descriptively as the time a girl could expect to have her first menstrual period. This is relevant because one of the markers of cultural difference is the different contemporary terminology and language, which reflects cultural beliefs about what it describes.

Menarche

The idea that the onset of menstruation is a mark of change of a girl's condition, of course, is one which symbolically still holds for many women today. Early modern medical texts had a broad consensus that a girl could expect to become menstruant between the ages of 12 and 14. There are no records that I have yet been able to find of young women openly recording their menarche in their diaries and/or letters, but perhaps one day an overlooked diary or set of correspondence might be discovered in the archives. This change of 'condition', as it was often designated, was a significant one, and it was even held that a young woman was now marriageable, but it was not the case that she was thought fully mature, and instead her life stage was seen as adolescent for a number of years more.

The sort of thinking which informed people at this time is seen in the theory of the seven ages of man, as vocalised by Jacques in William Shakespeare's early seventeenth-century play *As You Like It* :

> All the world's a stage,
> And all the men and women merely players;
> They have their exits and their entrances;
> And one man in his time plays many parts,
> His acts being seven ages. At first the infant,
> Mewling and puking in the nurse's arms;
> And then the whining school-boy, with his satchel
> And shining morning face, creeping like snail
> Unwillingly to school. And then the lover,
> Sighing like furnace, with a woeful ballad
> Made to his mistress' eyebrow.

(II.7)

Jacques describes men and women as players, but then goes on to describe a male lifecycle. The same incremental stages applied to women in this era too, however, as the opening suggests, and each of the seven ages describes a time period for each life-stage. So, while menarche might mark the moment a girl transitioned to young womanhood, her maturation process would continue, and part of that is, of course, exploring her sexuality and sexual urges.

The anonymously published *Aristoteles Master-piece, or, The secrets of generation displayed in all the parts thereof,* which appeared in many editions from 1684 right until the early part of the twentieth century, offered an explanation of the link between menarche and burgeoning sexuality:

> The propension and inclination of Maids to Marriage, is to be discovered by many Symptoms; as, when Nature fringes the obscure parts, and their Terms flow at the time appointed, which is usually in the Fourteenth or Fifteenth Year of their Age, when as the Seed increaseth in some sooner, and in others later, according to their Habits or Constitutions: And the Blood, which is no longer taken to augment their Bodies, abounding, incites their Minds and Imaginations to Venery.

(Anon. 1684, 5–6)

It explains that as girls have finished growing, the extra blood that was consumed in growth is now redundant and can cause their bodies to become overly hot and to experience sexual desire for the first time. In the humoral bodily economy, women's bodies were considered to be naturally colder and wetter than the male counterpart, and so this

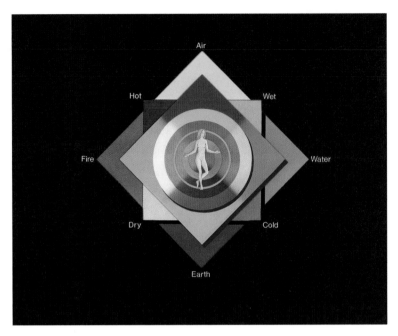

FIGURE 2.1 *The Four Elements, Four Qualities, Four Humours, Four Seasons, and Four Ages of Man.* Airbrush by Lois Hague, 1991. Wellcome Collection (CC BY-NC 4.0).

temporary accumulation of blood was unnatural (Figure 2.1). Men's bodies, it was held, were naturally hotter and used up their surfeit blood in exertion and in activities like growing facial hair, a secondary sexual characteristic linked to waste heat and blood; this meant that, prior to 1750 at least, some medics described beard growth as a kind of male menstruation (Withey 2020, 47–48). The explanation from *Aristoteles Master-piece* confirms that, just as Shakespeare had suggested, while boys in their third age were thought to begin dreaming of love and lovers, the same was thought true of girls too. And part of the physical changes which signify this is the onset of menstruation 'when they begin to be in Love, and desirous of Copulation; as also their short Breathings, Tremblings, and Pantings of the Heart', which in turn cause 'The Expulsive Faculties' to move 'to cast out the superfluous Humour' (Anon. 1684, 6) of blood.

It was not just in medical and pseudo-medical discourse that we find the idea that a young woman would be menstruating and therefore

capable of reproduction by age of fourteen. It had such cultural significance that the audience of another of William Shakespeare's plays,
The Winter's Tale, would know why Antigonus says he will geld his
daughters before they reach age 14 if Queen Hermione has been unfaith-
ful, to stop them behaving in the like sexually promiscuous manner. Of
course, Antigonus is playing devil's advocate here, trying to persuade
King Leontes of his wife's innocence, which is evident to all but he:

> I have three daughters; the eldest is eleven
> The second and the third, nine, and some five;
> If this prove true, they'll pay for't:
> by mine honour,
> I'll geld 'em all; fourteen they shall not see,
> To bring false generations: they are co-heirs;
> And I had rather glib myself than they
> Should not produce fair issue.

(4.2)

Another of the reasons this idea is so embedded culturally is that it fitted
with another theory, the belief that held that each seven-year anniversary
was a moment of potential crises in the body. Jane Sharp, author of *The
Midwives Book* published in 1671, explained that

> every climatericall or seven years breeding a new alteration in the body of
> Man: Children cast their Teeth at seven, and Maids courses begin to flow
> at fourteen. Seven times seven is of great danger to Mans life; and the
> great Climaterical which few escape is seven times nine, which makes
> sixty three.

(Sharp 1671, 174)

These points of change as set out in Jacques's speech were, then, antici-
pated as a moment of potential crisis for the body, one which heralded the
transition to a new life-stage at regular intervals into old age.

Menstruation

After menarche the norm was that a young woman might experience a
number of years of menstrual cycles, before marriage and children
followed for the majority. Menstruation was a normal and desirable part

of life for women in this era. It was thought key to keeping a woman's body in balance humorally, and was considered essential for demonstrating fertility. Yet, in the main, women, young or old, seem to have been disinclined to write about their cycles, despite the existence of the many, often quite vague, euphemisms for menstruation available to them. Katie Woodhouse-Skinner (2021) has conducted an extensive search of extant adolescent writings during the years 1660–1785 and has found one example of a young woman recording her cycle in her private diary. Sixteen-year-old Frances Sneyd marked twelve entries of the letter 'X' in her diary at monthly intervals.[1] As rare as it is to see such notation, it could be that women more broadly did mark their almanacs or other records in this way but that the private writings have not survived. However, Jane Sharp, the author of a guide to reproduction for midwives and lay readers alike, laments that

> Young women especially of their first Child, are so ignorant commonly, that they cannot tell whether they have conceived or not, and not one of twenty almost keeps a just account, else they would be better provided against the time of their lying in, and not so suddenly be surprised as many of them are.
>
> (Sharp 1671, 102)

This comment might allow the reader to infer either that older women did keep a proper record of their menstrual cycle, either formally or not, or, which is perhaps more likely, that older women had experience of recognising the signs of pregnancy and labour early and anticipating a potential pregnancy and subsequent delivery.

It has been possible to map some women's writing about their menstrual cycle from women who have had problematical experiences. These tend to be women from the upper ranks simply because more of their writings are extant. For example, Lady Margaret Hoby's diary and the letters of Brilliana, Lady Harley both refer to times when they were

---

[1] Katherine Woodhouse-Skinner, unpublished PhD thesis, Loughborough University, 2021. Warwickshire Country Record Office, 'The CR136 Diaries of Francis Parker Newdigate (1751–1835) and his wife Fanny [née Francis Sneyd], eldest daughter of Ralph Sneyd of Keele Hall, Staffs; she married Francis Newdigate (1751–1835)'.

unwell and in need of medical assistance in ways that suggest they were suffering from problems related to their menstrual cycles.

While women did not seem inclined to record their cycles even within their private diaries unless they were in ill health, their husbands did not feel the same, and there are a number of cases of men noting their wife's menstrual periods which exist today. In the sixteenth century, the alchemist and astrologer John Dee even makes discursive comments about the nature of his wife's bleeding, describing how the 'show' was 'abundant' or 'small' depending on how heavy he judged her bleeding to be (Woolley 2002, 283). Diarist Samuel Pepys is famous for recording his wife's terribly painful periods and records several times when she was really quite poorly. Indeed, the first entry in his diary of life in London, written on 1 January 1660, noted that Elizabeth and he had thought she might be pregnant but that the return of her 'terms' after seven weeks had ended their hopes. Despite the fact that the Pepyses had a stormy marriage, there are some quite touching moments recorded in the diary, such as when, on 9 May 1661, Pepys frets about workmen making a mess in his house while his wife is in bed, she 'being ill and in great pain with her old pain, which troubled me much because that my house is in this condition of dirt'. Frustratingly, Pepys does not record what steps Elizabeth took to alleviate her symptoms, but they would have been likely to focus on purges such as laxatives and emetics, since rebalancing the humours was thought key to easing dysmenorrhoea.

Why Menstruation Occurred

Nowadays menstruation is understood to be the loss of the womb lining and blood and other matter, whereas in the past menstruation was thought to be made up of blood lost from uterine veins. There were three main theories which offered reasons why a woman experienced a monthly bleed. Galen, writing in the second century CE, suggested that it was necessary for women to bleed each month because they led sedentary lives and so built up an excess of blood. This was unique to the female body, as men were assumed to work harder and engage in more physical exertion and so sweat out their excess humours, or, as we have seen, used them in the production of secondary male sexual characteristics such as beards. Galen's views, which were informed by the Hippocratic corpus, prevailed until the era under consideration today, and so Helkiah Crooke, physician to King James I, could write in 1615 that women were known to 'live an idle and sedentarie life, pricking for the most part vppon a clout' (Crooke 1615, 274), or sitting doing embroidery. Of course, Crooke's assertion is based on the lifestyle of elite women, and the many women who worked in often strenuous occupations and ran busy homes would not recognise this characterisation. In this model, though, the relative inactivity meant that over the month an excess of blood built up until it was discharged as a menstrual period. This explanation for menstruation was known as the plethora theory and was the most common explanation for the reason for women's periods, largely owing to the great importance placed on evacuations in humoral medicine. Physicians regularly prescribed medicines to make their patients vomit or go to the toilet and then gleefully recorded how many times a patient had been to 'stool' after a treatment. This was because evacuation in all forms was thought to be one of the environmental factors in maintaining health. That is to say, getting unnecessary matter out of the body was one of the six non-natural factors thought to affect bodily functions: the air one breathes, sleep, intake (food and drink), evacuations (including sexual emissions), movement, and emotions. In this context, then, it is easy to see why maintaining regular menstrual periods was so important to early modern women and their physicians.

The two other theories put forward about why women had periods were known as the ferment and lunar theories. The latter quite straight-forwardly held that the action of the moon caused women to bleed on a monthly basis, but, although it was being put forward in print as late as the 1970s, early modern medical writers regularly wrote that this theory was so flawed as to be easily dismissed. The main reason given was that, if the moon caused periods, then every woman of menstruating age would have her menstrual period at the same time as each other.

The lunar and ferment theories came together in Richard Bunworth's *The Female Doctoress* (Bunworth 1656), which explained how the moist, cool female body was dictated to by its 'governesse, the Moon' (Bunworth 1656, 2). The text explains that 'the humours of a Womans body have, once every month, their spring tide, not onely fermenting within their usuall limits, but also swelling to a greater extent, and are extravasated into the wombe to fertilize the same'. In this text, the notion that, if the moon determined a woman's cycle, then all women would expect to bleed at the same point, was addressed in the following way:

> Experience teacheth that young Women have their monthly evacuation [when] the Sun and Moon are in Conjunction, that is, at the time of New Moon: and ancient Women, who have still their months, have them when the Moon is opposite to the Sun, that is, at the time of Full Moon. Other Women, according as they are more or lesse in years do symphathize with the proportionable age of the Moon, and are purged at other intermediate Aspects: Viz: Sextile, Trine, Quartile, &c.

Most medics would concede the latter point that the moon did have some influence on women's cycles such that younger women were more likely to bleed at the rising moon and older ones as the moon waned. The ferment theory was fashionable for only a few decades. It held that, much like in the fermenting of alcohol, somewhere in the female body, a ferment brewed until it reached a critical point, and then bleeding broke out, and so the process began again every month. In 1707 James Drake's *Anthropologia Nova; or A New System of Anatomy* proposed that it was the function of the gallbladder to produce this ferment, but his hypothesis was not taken up, and with no evidence as to the site of this ferment being found, this theory died out (King 2009, 62).

Defloration: The Second Transition

After becoming menstruant, the second of the three transitional bleedings a female body was understood to undergo traditionally happened on her wedding night, when a maid became a wife. The blood lost on first intercourse is arguably imbued with more cultural significance than the other occasions I discuss. The fact that this blood attracted so much attention in the medical texts is largely because of the unique theoretical importance it carried, not only in acting as proof positive of virginity upon marriage but also because it was a second occasion of bleeding, following menarche, which formed part of a woman's growth to maturity. This is another one of the instabilities of the significance of this blood, however. Almost all medical texts take pains to offer reasons as to why a woman might not bleed on first intercourse, which included if she had a history of heavy menstrual periods which might have eroded the hymen and or loosened the passages. Given that very many women were not virgins on marriage, and indeed between a fifth and up to a third of them were pregnant at the time of their wedding (Hair 1966) – as was William Shakespeare's own bride Anne Hathaway, of course – and that this expectation carried a great deal of cultural significance for something which might prove elusive at the required moment, it required immediate plausible explanation.

To understand the relationship between this transition and menstruation, it is important here to see how the anatomy of the vagina was envisioned and described. In the 1663 text *Bartholinus Anatomy* a paradigmatic description of the vaginal anatomy is offered:

> In the Bottom of the Womb we have observed three things; the Bottom it self, the lesser Neck, and the Orifice. In the greater Neck also, three things are to be noted. The Neck it self, the Hymen, and the Mouth of the Bladder.
>
> (Bartholin 1663, 72)

In common with many medical texts, this one uses the term 'neck' to refer to the vagina. The term vagina itself only appears in English texts from 1612, when it is used in an English translation of Jacques Guillemeau's *Child-birth; or The Happy Deliverie of Women,* but is not widely used until the end of this century, next appearing in print in Thomas Gibson's *The Anatomy of Humane Bodies Epitomized* (Gibson 1682). The blood supply to this area is explained by Helkiah Crooke, who wrote that

> there are two veines which disperse their branches through the wombe, some of which are carried to the inward cavity thereof by which the infant is nourished, others run to the outward part of the wombe even unto the necke and the lap it selfe.
>
> (Crooke 1615, 315)

This understanding of female anatomy is further exemplified in the middle of the seventeenth century by the author of *The Compleat*

Midwives Practice, who states that the reason that there are so many veins in the vagina is that 'the flowers [menstrual periods] must not onely come out of the womb, but out of the neck of the womb also' (Anon. 1656, 29). The neck of the womb could mean the cervix, as it tends to do nowadays, but also as here it means the vagina, and many medics considered that the blood came from the veins of the vagina being stretched and so discharging blood on first intercourse. As Thomas Gibson explained in his 1682 anatomy guide:

> In Virgins its [the vagina's] duct is so straight, that at their first congress with a Man they have commonly more pain than pleasure through the extension of it by the Penis, whereby some small Vessels break, out of which Bloud issues as out of a slain Victim.
>
> (Gibson 1682, 154)

This last remark also clarifies that this blood was seen as a form of menstrual bleeding since as far back as the Hippocratic corpus, in which healthy menstrual blood was described as resembling the blood of a sacrificial victim. This connection is why intercourse, for example, was thought to be a dependable cure for the so-called virgin's disease, greensickness, an illness characterised by a lack of menstruation in young women. Lazare Rivière, Professor of Medicine at the University of Montpellier, explained this connection by saying that 'Experience teacheth, that sometimes these Women [with greensickness] have their Terms the first night after marriage, and that others who are in good health have them before their accustomed time' (Rivière 1655, 403). Greensickness was characterised by a lack of menstruation and loss of appetite. A resumption, or a start, of menstruation both facilitated recovery and confirmed that the patient was returning to health. The fact that intercourse was a key treatment for this disease suggests that, once the vaginal veins are opened during first coitus, they would then begin to function normally, and a woman would remain menstruant thereafter.

That intercourse caused these blood vessels to open, and bleed, is the explanation behind a piece of gossip that Samuel Pepys diarised on 8 January 1666, which was put out to provide a cover for one of the Duke of York's many affairs. The entry states:

> Sir Rd. Ford did this evening at Sir W. Batten's tell us that upon opening
> the body of my Lady Denham, it is said that they found a vessel about her
> matrix, which had never been broke by her husband, that caused all pains
> in her body – which, if true, is excellent invention to clear both the
> Duchesse from poison and the Duke from lying with her.

Lady Margaret Denham had been married the previous year at the age of
23, to the 50-year-old, apparently impotent, Lord John Denham. The
ruse that she died a virgin of a complication caused by a blocked vessel,
perhaps envisaged as a form of greensickness, was put about to remove
the supposition that she had been poisoned for being a favourite mistress
who held too much sway over the Duke of York (Ó Hehir 1968, 180).
That this story could be spread plausibly succinctly demonstrates that
there was a widespread acceptance in early modern society that the
vessels in the vagina contributed to menstrual bleeding and were indeed
a source of the blood associated with defloration.

Some commentators added that they thought this blood might eman-
ate from more than one source. Thomas Gibson, cited above, wrote that
he supposes the blood to come from the vaginal vessels 'unless we should
rather think that the Bloud proceeds from the rupture of the Hymen'
(Gibson 1682, 154). The hymen was described by Gibson as being

> a thin Nervous membrane interwoven with carnous [fleshy] Fibres, and
> endowed with many little Arteries and Veins, spread across the duct of the
> Vagina, behind the insertion of the neck of the Bladder, with a hole in the
> midst that will admit the top of ones little finger, by which the Menses
> flow.
>
> (Gibson 1682, 154)

This portrayal of the intact hymen being big enough to admit a man's
finger remains fairly consistent throughout the period. Descriptions of
the hymen normally compared it to a flower, often a carnation. Crooke's
account is fairly typical: he says it looks like 'the cup of a little rose halfe
blowne when the bearded leaves are taken away' or 'the great Clove
Gilly-flower when it is moderately blown' (Crooke 1615, 235). Jane Sharp
echoes this description almost word for word, but elaborates to explain
that 'thence came the term deflowered' (Sharp 1671, 48). However, the
very existence or not of the hymen was also heavily theorised in the

period, which is one of the reasons why there was a need for an alternative explanation for this issue of blood.

The volume of blood medical authors suggested a woman could expect to lose on this occasion was varied. According to medical understandings, the amount of blood lost could vary from an insignificant smear to life-threatening amounts. Dutch physician Ysband van Diemerbroeck recalled the death of one of his patients on her wedding night from haemorrhage:

> I remember, that I knew a young Bride in upper Batavia, to whom, by the violent immission of the Yard in the first Act of Coition, and suddain dilatation of the Vagina, there happen'd such a prodigious Flux of Blood, that in three hours she lost her Life, together with her Virginity. And the like unfortunate Accident some years ago befell the Daughter of a certain Citizen of Utrecht, who was so wounded the first night, that before morning, the Flux of Blood not being to be stopp'd, she expir'd.
>
> (van Diemerbroeck 1694, 177)

Additionally, while its appearance was proof of virginity, its absence did not disprove the same. Yet, following the lead of the ancient Greek physicians, in the early modern period it was considered necessary for a girl to lose her virginity for her to become a woman, and a loss of blood was proof of this transition and so was thought desirable.

Postpartum Blood loss

After defloration, pregnancy was expected to follow, and whereas I have not located any instances in life-writing of women discussing hymenal bleeding, they do occasionally record their postpartum bleeding. As previously discussed, postpartum bleeding was considered to be a form of menstrual bleeding. So, if a woman began bleeding in pregnancy, then many medical books would refer to this too as an attack of the menses. *Lochia*, the term now used for postpartum bleeding, was a term that was naturalised into English from the Greek in the early seventeenth century, but it tended to be used only by elite male writers rather than women. Jane Sharp does not use lochia, for example, but uses 'purgations' and menstrual blood, and fellow midwife Sarah Stone also does not refer to lochia, but uses 'cleanings' or 'flooding' for regular and heavy postpartum bleeding, respectively. Stone, whose midwifery memoir *A Complete Practice of Midwifery* was published in 1737, offered several examples of women who had problems with their 'menses' while pregnant, and she claimed to have a secret method to stem this bleeding which never failed, but which she declined to divulge, saying that she had to think of her daughter who was now a midwife, and so could not reveal this secret life-saving method in her memoir. Stone wrote that

> It is a secret I would willingly have made known, for the benefit of my Sisters in the Profession: But having a Daughter that has practised the same Art these ten years, with as good success as my self, I shall leave it in her power to make it known.
>
> (Stone 1737, 148)

It was widely believed in the seventeenth century that the foetus was nourished by menstrual blood. This, then, was given as an explanation as to why women sometimes experience bleeding resembling menstrual periods in early pregnancy. The foetus could simply not consume all the blood lost in a typical menstrual period while tiny in the early months; after this, it was thought that the baby took the choicest part of the blood and left the rest. This stored menstrual blood was what was thought to be discharged after the birth. The length of time that women were expected to bleed for similarly matched the time for which women had not bled whilst pregnant. Jacques Guillemeau explained that:

> This time may bee also measured according to that ordinary time of purging, that is omitted in the nine moneths, she goes with child, as the bloud should bee purged in euery one of these nine moneths, as in euery one of them, the space of three or foure daies (which put together amount to twenty seuen or thirty six dayes) so in recompense heerof, when a woman is deliuered, she must bee purged, 27. or 36. daies.
>
> (Guillemeau 1612, 221)

Yorkshire gentlewoman Alice Thornton wrote about her lochial flow as 'those' and the heavy bleeding she experiences after some pregnancies as 'floodings' in her autobiographical writings. After the birth of her daughter Betty, Thornton described how she was 'at death's dore by the extreame excesse of those, upon the fright and terror came upon me, soe great floods that I was spent, and my breath lost, my strength departed from me, and I could not speake for faintings' (Jackson 1875, 96–97). However, Thornton's frankness is extremely untypical. Thornton was an exceptional woman, writing in specific circumstances to shape her legacy. But her writing offers an insight into the ways that women might have routinely discussed their lochial flow with their midwife or their companions, as she uses the same term as Sarah Stone for excessive bleeding.

In this era, it was customary for a woman to remain in her bed after giving birth, rising incrementally, beginning with 'upsitting' after a few days and going on to pottering about her chamber before being able to resume her normal activities after a month. This period was often referred to as 'the woman's month', and is one of the reasons Hermione argues for the cruelty of being put on trial in *The Winter's Tale* before she has been allowed to complete her lying in:

> Proclaimed a strumpet: with immodest hatred
> The child-bed privilege denied, which'longs
> To women of all fashion.

> (3.2)

Hermione is right to suggest this was a privilege for women of fashion or quality, for while it was a recommendation for all women, for those whose families were economically dependent on their labour for survival it was not a luxury they could all afford. Midwife Sarah Stone records instances of laundry maids being back at work within three weeks.

One commonality that hymeneal and lochial bleeding share is the way they intertwine with religion. Whilst there was no ceremony associated with menarche, it did signal, as has been discussed, that a girl had become a young woman and was therefore marriageable. Defloration was implicitly – if unreliably – linked to the wedding ceremony, and so too was the occasion of postpartum bleeding something which was seen as a concern of the church. Following the end of the woman's month and the end of her postpartum bleeding, a woman was, for much of the early modern era, obliged to take part in the ceremony of Churching, in which thanksgiving prayers were offered for the woman's safe delivery.

Menopause

To complete the life cycle of female bleeding, it is useful to end with a few very brief comments about early modern attitudes to menopause. This was not quite conceptualised as a transition in the same way the other occasions of bleeding were, largely because it is marked by a gradual decline and cessation of bleeding. The first thing to note is that the term 'menopause' did not appear in print until 1852, and then in an article for *The Lancet* which suggests the term had some currency in medical circles, if not among the general populace. Before this, physicians used to refer to the cessation of the menses.

Physician John Freind, author of the first monograph on the topic of menstruation in England, first published in Latin in 1703 (Figure 2.2), wrote that:

> The menstruous Purgation, or a flux of Blood issuing from the Uterus every Month, usually begins its Periods at the Second Septenary, and terminates at the Seventh, or the Square of the number seven.
>
> (Freind 1729, 1)

So, we have come full circle and have arrived back at the seven ages of man theory, and many texts give age 49 as the natural ending of menstruation. Given that it too, like menarche, was a bodily change associated with a climatical year, it was one which some women might approach warily. Indeed, some medics argued that the way a woman experienced her menopause would set the tone for her health for the rest of her life. That this was known to be the end of a woman's reproductively active years is shown in comments such as in *Aristoteles Masterpiece*, which reads:

> after it cease to be with her after the Custom of Women, that is her Courses are stayed, which in some happens sooner and some later, and between 44 and 55 with them all [...] those Women must despair of further Generation: for as those learned in this Art frequently observe, where there is neither Buds not Blossoms there can be no Fruit.
>
> (Anon. 1684, 83–84)

FIGURE 2.2 John Freind. Line engraving by G. Vertue, 1730, after
M. Dahl, 1725. Wellcome Collection (public domain mark).

This text uses the fruit and flowers horticultural metaphor as it was
the one which would be most familiar to the text's general readership.
How far and in what ways women suffered from the symptoms that we
now associate with the menopause is not yet known, but symptoms such
as 'flashing' or hot flushes were discussed in medical books. Mostly
though, in personal writings it seems that women associated the menopause with the end of fertility, and fertility, as we know, diminishes
gradually in the run up to menopause, and this could be unwelcome to

some if they still had hopes of more family, but equally might well have been a relief to those women who had spent much of their married life pregnant.

I have been working on attitudes and accounts of women's reproductive bleeding for more than 15 years now, through from my MA and doctoral research and beyond. It is an area in which we still have much to discover. One of the things which adds to my excitement and enthusiasm to continue researching in this medical humanities field is that new works, printed and manuscript sources, are waiting to be discovered. As one example, Alice Thornton's autobiography was printed in Victorian times, yet recently another manuscript iteration of this text has been rediscovered in the archive of Durham Cathedral by historian Dr Cordelia Beattie, which is now being studied for the insights it may offer into her life. I remain hopeful of finding more personal writings by women in the early modern era which will build on the observations I have presented in this essay.

My thanks once again to all at Darwin College for extending this invitation to share my research here today as part of a fascinating lecture series on the theme of blood.

References

Anon. (1656) *The Compleat Midwives Practice, in the Most Weighty and High Concernments of the Birth of Man.* London: Nathaniel Brooke.

Anon. (1684) *Aristoteles Master-piece, or, The secrets of generation displayed in all the parts thereof.* London, J. How.

Bartholin, T. (1663) *Bartholinus Anatomy Made from the Precepts of his Father, and from the Observations of all Modern Anatomists; Together with his Own,* translated by Nicholas Culpeper and Abdiah Cole. London: Peter Cole.

Bunworth, R. (1656) *The doctresse: a plain and easie method, of curing those diseases which are peculiar to women.* London: Nicolas Bourne.

Crawford, P. (1981) 'Attitudes to Menstruation in Seventeenth-Century England'. *Past and Present* 91, 47–73.

Crooke, H. (1615) *Mikrokosmographia, a description of the body of man.* London: William Jaggard.

Diemerbroeck, Y. van. (1694) *The Anatomy of Human Bodies*, translated by William Salmon. London: W. Whitwood.

Drake, J. (1707) *Anthropologia Nova, or a New System of Anatomy*. London: Samuel Smith and Benjamin Walford.

Freind, J. (1729) *Emmenologia*, translated by Thomas Dale. London: T. Cox.

Gibson, T. (1682) *The anatomy of humane bodies epitomized*. London: M. Flesher for T. Flesher.

Guillemeau, J. (1612) *Child-birth; or The Happy Deliverie of Women*. London: A. Hatfield.

Hair, P. E. H. (1966) 'Bridal Pregnancy in Rural England in Earlier Centuries'. *Population Studies* 20(2), 233–243.

Jackson, C. (ed). (1875) *The Autobiography of Alice Thornton*. London: Mitchell and Hughes.

King, H. (1998) *Hippocrates' Woman: Reading the Female Body in Ancient Greece*. London: Routledge.

King, H. (2009) *The Disease of Virgins: Green Sickness, Chlorosis and the Problems of Puberty*. London: Routledge.

Ó Hehir, B. (1968) *Harmony from Discords: A Life of Sir John Denham*. Los Angeles: University of California Press.

Porter, R. (2003) *Flesh in the Age of Reason: The Modern Foundations of Body and Soul*. London: W. W. Norton.

Read, S. (2013) *Menstruation and the Female Body in Early Modern England*. Basingstoke: Palgrave Macmillan.

Rivière, L. (1655) *The Practice of Physick in Sixteen Several Books*, translated by Nicholas Culpeper, Abdiah Cole, and William Rowland. London: Peter Cole.

Sharp, J. (1671) *The Midwives Book, or, The whole art of midwifry discovered*. London: Simon Miller.

Stone, S. (1737) *A Complete Practice of Midwifery. Consisting of Upwards Forty Cases or Observations in that valuable Art, selected from many Others, in the Course of an Extensive Practice*. London: T. Cooper.

Withey, A. (2020) *Concerning Beards: Facial Hair, Health and Practice in England, 1650–1900*. London: Bloomsbury.

Woodhouse-Skinner, K. (2021) *Recovering Female Adolescence in Adolescent Life Writing and Socio-medical Discourse in England between 1660 and 1785*. PhD thesis, Loughborough University.

Woolley, B. (2002) *The Queen's Conjuror: The Life and Magic of Dr Dee*. London: Flamingo.

3 Blood in Motion, or the Physics of Blood Flow

TIM PEDLEY

This chapter starts with a brief survey of ancient ideas about blood flow, culminating in the West with William Harvey's convincing demonstration – before the invention of the microscope – that the blood circulates in the body. The mechanics of that circulation will be described, from the high-pressure arteries to the low-pressure veins. Highlights will be the following: the propagation of the pressure pulse in arteries, the disturbance to smooth flow caused by the complex geometry of arteries (and its probable influence on arterial disease), the fact that blood cells have to be deformed and travel in single file in the smallest capillaries, and the interesting effects of gravity on the venous return to the heart in upright animals, notably those with long necks and legs – giraffes and dinosaurs.

These days we all know that blood circulates, driven by the beating of the heart, in two distinct circulatory loops: the systemic circulation, through most of the tissues of the body, and the pulmonary circulation, through the lungs (Figure 3.1). The former, powered by the left ventricle of the heart, supplies oxygen and nutrients via a branching system of arteries to the capillaries in the active organs and tissues, which convert them into energy and proteins and produce metabolites, especially carbon dioxide (CO_2). From the capillaries the blood enters a converging system of veins, of which the largest, the *vena cava*, delivers the blood to the right atrium of the heart. Blood flow to the lungs is powered by the right ventricle; the lungs take the deoxygenated blood, remove excess CO_2, top up the oxygen supply and return it to the heart via the left atrium. Backflow to and from the four chambers of the heart is prevented by four non-return valves. A typical blood cell takes about a minute to go round the system once. However, it was not until the early seventeenth century that physicians, scientists, and scholars in the West (i.e., Europe)

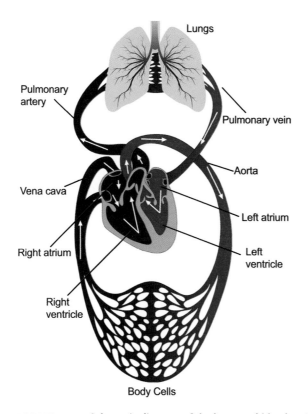

FIGURE 3.1 Schematic diagram of the heart and blood vessels in the systemic circulation (below) and the pulmonary circulation (above). Oxygenated blood is shown as red; deoxygenated blood as blue.

learnt about the circulation through the insight of William Harvey (1578–1657).

Old Ideas

The medical education that Harvey and his contemporaries received based its teaching on the work of the ancient Greeks, culminating in the prolific writings of Galen (129–199 AD/CE). Galen was a great scientist who recorded an enormous amount of information on human anatomy and physiology, in particular concerning the heart, blood vessels, and blood flow. He knew about the four heart valves, and that they prevent backflow, but, because he could not see the vessels of the microcirculation, he did not think of the blood as circulating constantly (though he did reckon, wrongly, that there might be very small openings between the ventricles and atria). The general idea was that the blood would travel out and back along the major arteries and veins, which were clearly distinguishable because of the different structure of their walls – thick or thin – and different appearance of the blood. The driving force was supposed to be the squeezing and relaxing of the central vessels as a result of the contraction and expansion of the thorax during breathing. The purpose of the blood flow remained rather vague – to convey nourishment, air in some form (pneuma), and, especially, heat between the major organs, e.g., heart, liver etc., in order to maintain the balance of the Hippocratic 'humours'.

In fact, unknown to Galen, the idea that blood circulates had been conceived in ancient China: the *Huángdì Nèijīng* (黄帝内经) (third century BC/BCE) says that 'the blood current flows continuously in a circle and never stops. It may be compared to a circle without beginning or end.' Also in the third century BC, Wáng Shūhé wrote the *Pulse Classic* (*Mài jīng*, 脉经), and innumerable commentaries were written on his work. The pulse is examined in several places, at different times, with two or three fingers simultaneously and with varying degrees of pressure. The process may take as long as three hours. It is often the only examination made, and it is used both for diagnosis and for prognosis. Not only are the diseased organs ascertained, but the time of death or recovery may be foretold. I am unaware of any successful test of this system using modern quantitative methods.

Another important centre of medical science was Cairo, in the twelfth and thirteenth centuries AD, the 'Golden Age of Arabic science', where first Maimonides (1135–1204) (a Jewish rabbi) and later Ibn al-Nafis (1213–1288) proposed that blood must pass from arteries to veins without returning to the heart, in the systemic and pulmonary systems, respectively. Ibn al-Nafis was explicit in asserting that all the blood that reached the left ventricle had passed through the lung and that there must be small communications between the pulmonary artery and vein, a prediction that preceded by 400 years the discovery of the pulmonary capillaries by Marcello Malpighi (Rosen 1995; West 2008). However, this work, like that of the ancient Chinese, was not known in Europe.

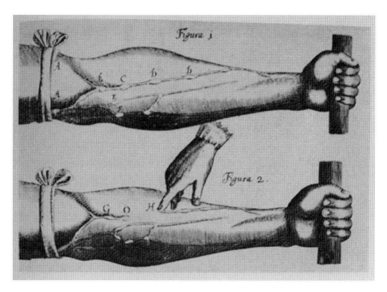

FIGURE 3.2 Harvey's demonstration of the presence of one-way valves in the veins of the arm. A not-very-tight ligature round the upper arm squeezes the veins shut but still allows arterial blood into the forearm. The veins become distended, and little swellings – C, D, D, etc. – indicate the presence of the valves. If you 'milk' the vein from O to H you can squeeze the blood out of that segment of the vein, and it does not fill up despite the clearly greater pressure to the left of O – the valve prevents backflow.

In Europe there were several scientists (I use the modern term) who performed careful anatomical and physiological observations of the heart and blood vessels; notable among them was Leonardo da Vinci, for example, with his detailed drawings of the heart and its valves. However, they did not challenge the Galenic view. That nettle was grasped by Harvey and promulgated in his ground-breaking work *Exercitatio anatomica de motu cordis et sanguinis in animalibus: Anatomical Treatise on the Motion of the Heart and Blood in Animals* (Harvey 1628). The two arguments that were most persuasive were as follows. First, the direct demonstration of non-return valves in the veins of the arms (Figure 3.2), which was inconsistent with the proposition that blood could flow in either direction in veins; it had to flow towards the heart. Second, a quantitative estimate of how much blood was received and thus ejected by the left ventricle each beat (the stroke volume). Harvey's estimate of stroke volume was approximately 1.5 fl.oz., or 45 ml, leading to a cardiac output of around 3 litres per minute, which is a bit less than half the currently accepted normal value. But even if his was a five-fold overestimate, said Harvey, the quantity of blood moving along the arteries or veins would be vastly too great for it to be supplied constantly by the 'juice of ingested food'.

Understanding the dynamics of blood flow – the forces driving and resisting the motion and the resulting flow patterns – came later. Important evidence was obtained by the Reverend Stephen Hales (1677–1761), who made direct measurements of the arterial blood pressure in several animals by puncturing a carotid artery, inserting a cannula connected to a long glass tube, held vertically, and showed that the blood rose to a height of over 8 feet (2.44 metres), in a horse, for example (an extremely grisly procedure, which would not have been acceptable to ethics committees in the twentieth century, let alone the twenty-first!). This is equivalent to 183 mmHg, comparable to a fairly high systolic pressure in people as well as other mammals. He commented that this pressure pulsed in time but did not fall close to zero until several heartbeats after the artery was punctured. It was also clear that venous pressure remained very much smaller than that in a large artery (witness the fact that arterial blood squirted out in a jet if an artery were punctured during clinical bloodletting, whereas it dribbled out from a vein).[1] Among other findings, Hales noted that the peripheral blood flow was less intermittent than the ejection from the heart, and that this was a consequence of the elasticity of the blood vessels (again, using modern terms).

Leonhard Euler (1707–1783), probably the greatest applied mathematician of the eighteenth century, pointed out that the pulse must *propagate* like a wave along the elastic vessels and correctly wrote down the equations governing that propagation, but, because of unrealistic assumptions about the properties of the vessel walls, he was unable to solve them. It fell to Thomas Young (1773–1829) to calculate the speed of propagation of the pulse wave as expounded in his Croonian Lecture to the Royal Society in 1808. It must be said that his calculation is very

---

[1] Withdrawal of blood from a vein was considered therapeutic, both for restoring the proper balance between the 'humours' and for cooling the blood. It presumably remained popular for centuries because of the temporary euphoria that it can generate.

difficult to follow there, since it was presented using rather obscure physical arguments rather than the equations. Quantitative details are given below, but here, as a mathematical fluid dynamicist, I cannot resist the following direct quotation: 'The inquiry, in what manner, and in what degree, the circulation of the blood depends on the muscular and elastic powers of the heart and of the arteries, supposing the nature of those powers to be known, must become simply a question belonging to the most refined departments of the theory of hydraulics' (Young 1809). It is to some of those departments that this lecture is primarily devoted.

Dynamics of Blood Flow

Pulse Propagation in Arteries

Consider a single heartbeat. The left ventricle starts to contract as soon as it has been refilled from the left atrium and the mitral valve between them has closed. When the left ventricular pressure exceeds the aortic pressure, the aortic valve opens and the majority of the contents of the left ventricle are ejected into the aorta, where the pressure therefore rises and the aorta, being an elastic tube, expands locally. However, the blood is moving and starts to push the blood already present further downstream; this tends to reduce the excess pressure in the entrance of the aorta, so the elastic vessel wall tends to spring back, further increasing the speed of the blood just downstream and tending to inflate the vessel there. This process can be thought of as repeating itself in every subsequent segment of the aorta and propagates as a wave.

Figure 3.3 shows graphs of the pressure in a canine aorta measured as a function of time (Olson 1968). The different curves were measured at successively greater distances from the aortic valve; the pressure was measured with a transducer at the tip of a catheter that had been inserted in the aorta from the femoral artery in a hind leg. For each curve the catheter was withdrawn by 4 cm and a new curve plotted during a later heartbeat; the zero of time in each case was taken at a particular event on the electrocardiogram (ECG) that marks the opening of the aortic valve, assuming that the pressure pulse was unchanged over successive beats.

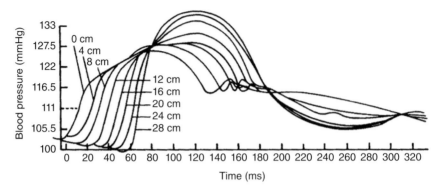

FIGURE 3.3 Plots of pressure in the aorta of a dog as a function of time after the opening of the aortic valve, at different distances from the valve, demonstrating the propagation of the pulse wave. After Olson (1968).

It can be seen from Figure 3.3 (i) that the pulse propagates, since the rise in pressure at the wavefront occurs later for more peripheral locations; (ii) the spacing between successive wavefronts decreases slightly as the distance from the heart increases, indicating that the speed of propagation increases with distance; (iii) the wavefront steepens a little as it propagates; and, most obviously, (iv) the wave height, or amplitude, increases dramatically as it propagates. A theory should be able to explain some or all of these observations.

Any mechanical wave (for example, a water wave), like any mechanical oscillator (for example, a pendulum), represents a balance between *inertia* and a *restoring force*. In the case of the pressure pulse, the inertia is that of the moving blood – its momentum – while the restoring force is provided by the elasticity of the vessel wall. The mathematical model initiated by Euler, using an equation that bears his name, considers wave propagation in a straight elastic tube of circular cross-section. The two governing equations represent conservation of mass (matter is not created or destroyed) and Newton's second law of motion, that if an element of the fluid is acted on by a net force F it will experience an acceleration a according to its mass m ($F = ma$). It is beyond the scope of this lecture to go into more detail, but, on the basis of a few more assumptions (blood is incompressible, the change in vessel diameter is much smaller than the undisturbed diameter, the blood velocity is much smaller than the wave propagation speed, and the blood's viscosity can be neglected), the model predicts that the speed of wave propagation c, in a tube of diameter D, wall thickness h, Young's modulus of elasticity (yes, the same Young) E, and blood density ρ, is given by

$$c^2 = \frac{Eh}{\rho D}. \tag{3.1}$$

If independently measured values are used for the parameters, this formula gives a prediction of c equal to about 5 m/s in the ascending aorta, rising to about 8 m/s in the abdominal aorta. These values are very close to those that are measured, for example from Figure 3.3. This is a remarkable achievement, that such a simple model gives results that are so close to observation. However, the model also predicts that the shape of the waveform of the cross-sectionally averaged blood velocity (the flow rate divided by the cross-sectional area) should be the same as that of the pressure wave form (Figure 3.3), which it is not, and that the shape of the wave does not change as it propagates, which it clearly does. The model must therefore be extended to incorporate some features that it does not yet contain.

An important feature that is ignored in the above model is the fact that no artery is long and uniform without interruptions: see Figure 3.4, which is a schematic diagram of the largest systemic arteries of a mammal (probably non-human since it appears to have a tail). Every artery has smaller branches; the aorta, in particular, has a major bifurcation below the abdomen, the iliac bifurcation into the two iliac arteries. Whenever a wave encounters an obstruction, it experiences complete or partial reflection. In this case some proportion of the pulse wave energy is reflected and the remainder transmitted to the daughter vessels. It turns out to be quite simple to analyse the reflection of a pressure wave at a bifurcation. We suppose that a known wave, of amplitude P_1, is incident on the bifurcation from the aorta and we wish to calculate the amplitudes of the reflected wave, P_R, and the two transmitted waves, P_2. The original wave theory tells us that the amplitudes of the corresponding flow-rate wave forms are $\Upsilon_1 P_1, -\Upsilon_1 P_R, \Upsilon_2 P_2$, respectively, where $\Upsilon_1 = A_1/\rho c_1$, A_1 is the cross-sectional area of the aorta (vessel 1) just above the bifurcation, c_1 is the wave speed in vessel 1 and ρ is again the

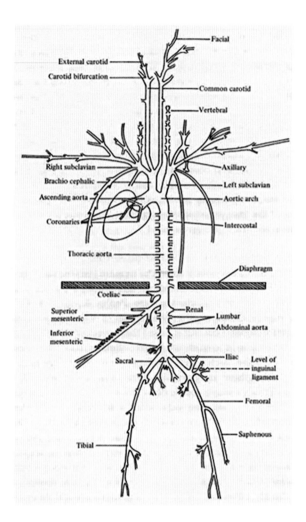

FIGURE 3.4 A diagrammatic representation of the major branches of the canine arterial tree. After McDonald (1974).

blood density; there is a similar definition of the quantity Υ_2 for the daughter vessels (suffix 2). The quantity Υ_1 is called the intrinsic *admittance* of vessel 1, and similarly Υ_2 for the daughter vessels (admittance is the inverse of *impedance*, which may be a more familiar term).

Two principles have to be applied to link the waves proximal and distal to the bifurcation. First, conservation of mass again – what flows into the bifurcation in tube 1 must flow out into tubes 2, at all times, which implies that $\Upsilon_1 P_1 - \Upsilon_1 P_R = 2\Upsilon_2 P_2$. Second, the total pressure must be continuous, for if it were not, a finite force would be experienced by an infinitesimally thin slice of the blood, which would therefore have an infinite acceleration. This condition gives $P_1 + P_R = P_2$. Combining these two equations leads to the prediction that

$$\frac{P_R}{P_1} = \frac{\Upsilon_1 - 2\Upsilon_2}{\Upsilon_1 + 2\Upsilon_2}, \tag{3.2}$$

as well as a prediction for P_2. Now, the total pressure amplitude at the bifurcation is $P_1 + P_R$, which is greater than P_1 if P_R is positive; in that case the pressure wave amplitude grows as it approaches the bifurcation from the direction of the heart; this is the peaking of the pulse, as seen in Figure 3.3. Equation (3.2) is consistent with this observation if $2\Upsilon_2$ is less than Υ_1. The wave speed c, from Equation (3.1), does not change sharply from aorta to iliac arteries, so the combined admittances of the daughter tubes will be less than that of the parent if the combined cross-sectional area is lower. In general, especially in the human circulation, the total cross-sectional area of the two iliac arteries is indeed normally smaller than that of the aorta, which is consistent with the observed peaking. Many other bifurcations appear to be well-matched ($\Upsilon_1 \approx 2\Upsilon_2$), so there is little reflection.

Thus, the simple model of Euler and Young, supplemented by the above representation of a bifurcation, is capable of explaining not only the speed of propagation of the pressure pulse but also its peaking. It also partly explains the difference in shape between the velocity wave form and the pressure wave form, though some of the details of that also involve viscous effects, and the steepening is a consequence of so-called non-linearity, which is significant when the average blood velocity is not so small compared with the wave speed (Pedley 2003).

Viscosity

The pulse wave theories of Euler and Young did not include the viscosity (frictional property) of the blood, although it was clear that blood, like any fluid, would experience friction as it is forced through a narrow tube

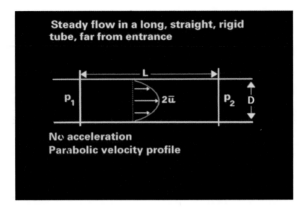

FIGURE 3.5 Steady flow in a long straight tube (Poiseuille flow) far from the entrance (see the text).

(driven *in vivo*, ultimately, by the pressure difference between arteries and veins). Jean Léonard Marie Poiseuille (1797–1869), a physician who wanted to know about the resistance to blood flow in small vessels, conducted a series of careful experiments in glass tubes using water as the flowing fluid (blood having proved unusable because of its unfortunate propensity to clot!). He found that the flow rate Q in a tube of diameter D and length L is proportional to the pressure difference between the ends (P_1-P_2) and to the fourth power of D, while being inversely proportional to L; Hagenbach (1860) later evaluated theoretically the constant of proportionality in terms of the fluid viscosity μ. The result is

$$Q = \frac{\pi(P_1-P_2)D^4}{128\,\mu L} \tag{3.3}$$

which is well known as *Poiseuille's law*.

Poiseuille's law applies to steady flow in a long, straight, rigid, circular tube, far from its ends; every element of fluid moves in a straight line at constant speed (Figure 3.5). The existence of viscosity (which Newton called the 'defect of slipperiness') stems from intermolecular forces in a continuous fluid; these also mean that there can be no slip between a fluid and a solid boundary. In fact, Poiseuille flow does not occur in any blood vessel! Why not? Consider the arteries (Figure 3.4). They are not long

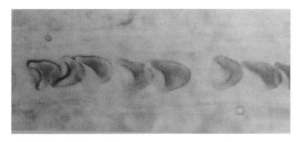

FIGURE 3.6 Red cells deformed in order to flow in single file along a narrow capillary. From Skalak and Brånemark (1969).

(between branches), nor straight, nor rigid; the geometry and the flow are not steady, because of the pulsed ejection from the heart. Moreover, blood is not a continuous liquid; it is a suspension of blood cells, mostly red blood cells, in plasma. The red cells are flexible biconcave discs, approximately 8 μm across, which occupy about 45 per cent of the fluid volume – this is so high that the blood would not flow at all if the cells were not highly deformable. In the smallest capillaries, 4–6 μm in diameter, the cells have to be highly deformed to fit in, and travel along in single file (Figure 3.6). Between the large arteries and the capillaries there is an extensive tree of increasingly numerous and increasingly narrow arteries, which distribute the blood to all parts of the body. The smallest arteries, whose walls contain muscle and are proportionally thicker than in large ones, merge into the arterioles of the microcirculation, where contraction of the muscle in the walls controls the local distribution of blood between different tissues, and eventually into the capillaries. Beyond the capillaries is the inverted tree of the venous system.

Inertia and Viscosity

Inertia is the property of matter to go on moving with the same speed, in the same direction, unless acted on by a net force (Newton's first law), when it experiences an acceleration and gains inertia (Newton's second law), depending on its mass ($F = ma$). In the case of liquids (almost-incompressible fluids) in tubes, the force consists of pressure gradients, that drive the flow and may accelerate it, and viscous forces, that resist fluid shear and dissipate energy. There may also be external forces such as gravity. The relative importance of inertia and viscosity is often estimated by calculating the *Reynolds number*, the ratio of a typical mass times acceleration ('inertial force') to a typical viscous force. For flow in a tube, this is

$$Re = \frac{\rho D \bar{u}}{\mu} \tag{3.4}$$

where the symbols have the same meanings as introduced above.

Table 3.1 lists various blood vessels, ranging in size from large arteries to capillaries, with their dimensions and typical velocities, and with the corresponding value of the Reynolds number Re (the data come from the canine circulation, which differs from the human case only insofar as the larger vessels are smaller in the dog; microcirculatory vessels are the same). To compute Re I have used a typical viscosity for whole blood of 0.004 Pa.s (although blood is a suspension of cells, it is a fact that, in blood vessels whose diameter is much larger than that of the cells, whole blood does behave as if it is a normal, continuous, 'Newtonian' fluid; this is a blessing for a theoretician like myself since it means that conventional fluid mechanics is applicable!). It can be seen that Re is much greater than 1 in large and medium-sized arteries, meaning that inertia is much more important than viscosity, which is the reason why the neglect of viscosity in the pulse wave analysis is a good approximation. In the microcirculation, on the other hand, Re is very much smaller than 1, meaning that inertia is unimportant there and the flow is dominated by viscous effects. In the microcirculation, of course, the blood cannot be considered as a continuous viscous fluid (Figure 3.6), but the conclusion that inertia is unimportant is not affected.

In the large arteries the mean pressure (neglecting gravity) remains more or less constant and, as we have seen, the pulse amplitude increases with distance. However, after several generations of branching, the vessel diameter and the flow velocity become small, viscous effects increase in importance, generating significant resistance, and both the mean pressure and the pulse amplitude decrease. In the microcirculation the pressure falls sharply and there is virtually no pulse, a state of affairs that persists into the venous system, apart from some small pulsations that propagate back from the right side of the heart.

Table 3.1. *Typical dimensions, velocities and Reynolds numbers in canine blood vessels (assuming blood viscosity = 4×10^{-3} Pa.s)*

Vessel	Diameter (mm)	Length (mm)	Peak velocity (m/s)	Mean velocity (m/s)	Re (peak)	Wave speed (m/s)
Ascending aorta	15	50	1.2	0.2	4500	5
Descending and abdominal aorta	13–9	350	1.05–0.55	0.2–0.15	3400–1250	8
Femoral artery	4	100	1	0.1	1000	8.5
Arteriole	0.05	1.5	0.0075	0.0075	0.09	–
Capillary	0.006	0.6	0.0007	0.0007	0.001	–
Venule	0.04	1.5	0.0035	0.0035	0.035	–
Vena cava	10	300	0.25	0.25	200	–

Should we conclude that, because of the high Reynolds number, viscosity is totally unimportant in the arteries? The answer is no, not because of its effect on pulse propagation, but because of the forces exerted by the flow on the artery wall. We recall the no-slip condition. This requires that, whenever fluid flows through a tube, there has to be a velocity profile between zero velocity at the wall and the throughflow within the vessel (Figure 3.5). Such shearing motion involves viscosity: whenever the fluid is sheared, i.e., there is a gradient of velocity across the flow, viscosity exerts an internal tangential force per unit area tending to reduce that gradient, called the *shear stress*. When the velocity gradient – the slope of the velocity profile – is α (measured in inverse time units) the magnitude of the shear stress is $\mu\alpha$, where μ is the viscosity. In the parallel flow depicted in Figure 3.5 the slope of the velocity profile is greatest at the wall and zero on the centre line). Hence the flowing blood always exerts a shear stress on the vessel walls – the wall shear stress (WSS) – and, because the walls are made of soft tissues, the WSS will tend to affect them in some way, at least by causing some deformation of the cells. This is thought to have an influence on the initiation (at least) of arterial disease: atherosclerosis.

Atherosclerosis and Wall Shear Stress

Atherosclerosis is the culmination of a disease process which in its early stages consists of the accumulation of excess lipid at the internal surface of the vessel wall, just beneath, or within, the endothelial cells (which form the inner lining of the blood vessels). As time passes, the lipid accumulation causes an increase in thickness and rigidity of the innermost layers of the wall. The accumulation continues to grow, becomes fibrous and turns into a thrombotic plaque, which, depending on its shape and where it is, may either obstruct the vessel so much that the flow is significantly reduced or break off and be swept downstream into peripheral vessels. Either way, the downstream tissues may be severely compromised; this is particularly serious if the plaque originated either in a coronary artery or in a carotid artery, causing a heart attack or stroke, respectively.

It has been observed, from post-mortem samples, that the accumulation of lipid begins quite early in the life of healthy human beings and is

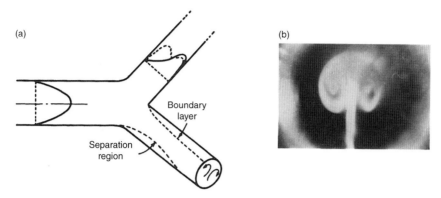

FIGURE 3.7 Steady flow in a symmetric bifurcation. (a) Note the high wall shear stress on the wall downstream of the flow divider (in the boundary layer), the low wall shear stress on the opposite wall where the flow might separate, and the generation of secondary motions because the flow follows a curved path. (b) Visualisation of the secondary flow in a daughter tube viewed from downstream; the fluid was air, visualised using smoke. From Schroter and Sudlow (1969).

normally clearly present in one's early twenties. It takes the form of fatty streaks, which have been visualised in other mammals (e.g., rabbits) by the enhanced uptake of certain dyes. Of interest here is the fact that fatty streaks are not uniformly deposited, but arise preferentially in large arteries, on the inside of curves and in the neighbourhood of branches. These are sites where the flow is markedly disturbed and therefore the distribution of WSS is highly non-uniform (see Figure 3.7). This observation led, in the late 1960s, to the hypothesis that the early development of atherosclerosis at a site is correlated with some property of the WSS experienced there. However, it was not clear what property; Fry (1968) proposed that a *high* WSS promotes the changes that lead to fatty streaks, whereas Caro et al. (1971) proposed that a *low* WSS is the key. This (perfectly amicable) disagreement triggered decades of clinical and bioengineering research to try to discover the mechanisms by which WSS might trigger the fatty streaking and, later, full-blown atherosclerosis. Such research has revealed an enormous amount about how the flow can affect the cells and tissues of the artery walls, but I believe it is still the case that no one has clearly understood the whole process. This is largely due to the enormous range of timescales involved, from fractions of a second (the timescale of a heartbeat) to the years it usually takes for atherosclerosis to develop. If anyone had been betting in 1970, I think the low-WSS hypothesis would have been then, and still is, the bookies' favourite over the high-WSS hypothesis, but many other runners have subsequently entered the race.

An initial idea of the complexity of flow in arteries can be obtained by considering a simple, symmetric bifurcation such as that used above to analyse pulse wave reflection, as drawn in Figure 3.7. We suppose there is a steady, parallel (Poiseuille) flow coming into the parent tube from the left, at a Reynolds number of a few hundred. The faster moving fluid in the centre will impinge on the flow divider and will be deflected into both of the daughter tubes, so the flow velocity will be maximum near that part of a daughter tube wall that extends directly downstream from the flow divider. However, the fluid is viscous and must be at rest at the wall, so there will be a thin region next to that wall, called a *boundary layer*, in which the velocity falls from its maximum to zero, resulting in a high WSS there. The velocity profile across the tube is sketched as a solid curve in the upper branch of Figure 3.7(a). On the opposite wall, the flow will have had to negotiate a convex corner, and, if that is sufficiently sharp, the faster flow will separate from the wall, leaving a zone of low velocities and low WSS beneath it; this is thought to be related to the greater incidence of fatty streaking in such locations. The separation zone and the boundary layer are sketched in the lower branch of Figure 3.7(a). The above description would be just as applicable to a two-dimensional geometry, in which the flow conduits were planar channels, as it is to the three-dimensional case considered here. The main difference in the three-dimensional case is that, as the flow into a daughter tube follows a curved path, the faster-moving fluid is flung towards the outside of the bend by its inertia ('centrifugal force') and will displace the slower-moving fluid near the flow-divider wall, which will therefore move round the inside of the tube towards the opposite wall. The consequence is that a transverse *secondary motion* will be set up, consisting of a pair of longitudinal vortices in each daughter tube, as also sketched in Figure 3.7(a). These vortices will themselves have an effect on the velocity profile in a plane perpendicular to that of the bifurcation, indicated by the dashed curve in the upper branch. Figure 3.7(b) is a photograph of the secondary flow, taken from downstream in a daughter tube, in which the fluid was air and the flow pattern was marked by smoke (the above considerations are all applicable to airflow in the lung as well as to blood flow in arteries; see Schroter and Sudlow (1969)).

It follows from the above that fluid flow in even a simple-looking geometry may be very complex and would be hard to measure accurately or calculate; predicting the WSS quantitatively requires considerable computation. Many experiments and much related modelling have been performed on steady and unsteady flow in tubes of all sorts – curved tubes, tubes with small or large side branches, branches at various angles, tubes with local contractions (stenoses), elastic tubes in which the cross-section is unsteady – leading to very complex flows and unexpected distributions of WSS (Pedley 1995; Ku 1997). If we were to seek greater realism than the bifurcation of Figure 3.7, we would have to consider bifurcations that are not symmetric, vessels that are curved and have imperfectly circular cross-sections, flow that is unsteady, with variable unsteadiness, and walls that are elastic and therefore moving. Moreover, every subject would be different; one person's vessel geometry, state of health (e.g., the extent of atheromatous plaques, if any), heart rate, and almost everything else will be different from another's. In recent years there has been a move towards personalised medicine, so systems have been developed to image a patient's arteries, using magnetic resonance imaging (MRI) for example, to digitise the image, to record the patient's pulsatile aortic blood velocity (measured using ultrasound or also with MRI) and then to put all that into a computer code to solve the full set of governing equations (the *Navier–Stokes* equations), that incorporate both inertia and viscosity to predict the time- and space-dependent WSS distribution. The reader will realise that this would be a tremendous undertaking, and, although it is in principle possible, it is not clear that it would lead to clinically valuable information in a timely manner. One problem is knowing the right question to ask, and hence knowing how fine the time and space resolution of the measurements and the computations need to be. Not really knowing how WSS governs the development of atherosclerosis is not a good basis for large expenditure on imaging and computing equipment. It has not led to a widely used bedside tool!

The Venous System: Blood Pressure and the Effect of Gravity

In the above discussion we have been assuming that the blood vessels have a circular cross-section. They are elastic tubes, but, as long as the internal pressure is significantly larger than the external pressure, the vessel wall will be distended and the cross-section will inevitably be approximately circular. The external pressure is that in the surrounding tissues, which is everywhere close to atmospheric, with relatively modest departures, for example in the chest, whose volume changes during the respiratory cycle.[2] The internal pressure has three components: atmospheric pressure, the pressure generated by the heart, which is what is usually called *blood pressure*, and a component due to gravity – hydrostatic pressure. The pressure in a fluid at rest decreases with height, so that over a height h the pressure falls by an amount $g\rho h$, where g is the acceleration due to gravity and ρ is the fluid density. In the human arterial system, the mean blood pressure is approximately 100 mmHg, i.e., 13.3 kPa, at the level of the heart (where it is usually measured, as the reader will have noticed, so that it does not matter whether you are standing up or lying down), while the hydrostatic pressure at the top of the head is no more than 5 kPa lower (assuming a vertical distance of 0.5 m), so the internal pressure is at least 8 kPa (60 mmHg) there. In the feet it is much higher when the subject is upright. Hence the *transmural pressure* (internal minus external) remains positive throughout the arteries and the assumption of circular cross-section is normally justified.

The situation is quite different in the venous system, where, at the level of the heart, the external pressure is only slightly above atmospheric and the mean blood pressure is also low, as we have seen. At the level of the feet, the internal, and therefore transmural, pressure of all vessels including the veins is large, so the vein wall will be distended, but how will the blood be pumped back to the heart? If the tubes were rigid, the incoming blood from the arteries would be pushed through the microcirculation and on through the veins. However, with distensible veins much of the

---

[2] Intrathoracic pressure normally varies at most between ± 10 cmH$_2$O, i.e., ± 1 kPa or ± 7.5 mmHg, during normal breathing.

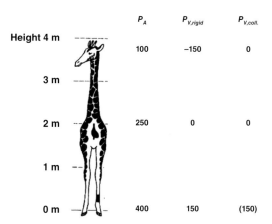

	P_A	$P_{V,rigid}$	$P_{V,coll.}$
Height 4 m	100	−150	0
3 m			
2 m	250	0	0
1 m			
0 m	400	150	(150)

FIGURE 3.8 Estimates of the approximate mean internal pressure in major blood vessels in an upright giraffe at different heights, assuming that the arterial pressure at the level of the heart is 250 mmHg. P_A is the arterial pressure, $P_{V,rigid}$ is the venous pressure if the vessels are assumed to be rigid, and $P_{V,coll}$ is the venous pressure if the veins are collapsible. All pressures are in units of mmHg. See Hargens et al. (1987). (100 mmHg is equivalent to 133 cm of water and hence of blood, or 13.3 kPa.)

flow could go towards filling them further. Moreover, there would be leakage through the capillary walls, which have to be permeable to water and solutes in order to supply the tissue cells, so the subject would be susceptible to a build-up of fluid in the tissues (oedema), leading to swollen ankles. This is where the venous valves come in: the blood is pumped not by the beating of the heart but by the contraction of the surrounding skeletal muscles squeezing the veins. As the subject contracts the muscles in the calf, for example, they squeeze the veins that pass through the muscle and push some of the blood up the leg, and this is prevented from falling back by the valves. The blood in a vein is pumped up, against gravity, in a series of steps, from one valve to the next, and so on. If you stand perfectly still for a long time you will indeed suffer from pooling of blood in the lower legs and the blood supply to the head may suffer; this is why guardsmen on parade are susceptible to fainting.

A different problem may afflict the blood vessels above the heart, and is especially acute in very tall subjects, such as giraffes. In a giraffe, the central arterial blood pressure at the level of the heart is 2½ times what it is in humans (and most other mammals) – see Figure 3.8. Why should this be and what are the consequences? The most obvious reason is that the internal pressure in the arteries in the head must be high enough to keep them open and to supply blood to the brain, whatever happens. If the blood vessels were rigid and totally impermeable, there would be nothing to stop the pressure in the vessels of the head from going negative and the flow rate being maintained, as in a siphon. This has been proposed in the physiological literature, but it does not explain how mass transfer to the tissues could then take place, and it fails to explain various features of the venous return (see below). In any case, a central arterial blood pressure of around 250 mmHg has been measured, although 200 mmHg is more usual.

The rate at which a pump does mechanical work, W – the power of the pump – is equal to the product of the mean pressure generated, P, relative to atmospheric, and the mean flow rate, Q, which can be expressed as the following equation:

$$W = (P - P_{atm})Q. \tag{3.5}$$

We may suppose that the mean flow rate per unit mass of tissue is approximately the same in giraffes as in people, so, if the central blood pressure is twice as great in the giraffe, the power required to pump the blood will also be twice as great, per unit mass of tissue. Hence there will need to be twice as much heart muscle, and indeed it has been found that the heart in an adult male giraffe is well over 1 per cent of body mass, compared with about 0.5 per cent in people. This is a consequence of having to pump the blood uphill by more than 2 metres in order to keep the brain perfused and is a significant cost. A corollary is that, if a dinosaur with a 15-metre neck raised its head vertically in order to feed, then it would have needed a heart of 7.5 per cent of body mass, which is highly improbable. Indeed, modern palaeontological research on the neck vertebrae of large dinosaurs suggests that their necks were unlikely to have been raised more than a few degrees above the horizontal.

Another cost relates to the need to prevent oedema in the lower leg, where the internal pressure will be 400 mmHg, but the giraffe does not have swollen ankles. There are two factors that help explain this: one is that the walls of capillaries in the leg of the giraffe are significantly less permeable than those in humans. The other is that the skin of the giraffe is much thicker and tighter than human skin and cannot be readily distended – there is nowhere for excess fluid to go except up. It is as if they wear anti-gravity stockings all the time.

A further question that arises is the venous return from the head, flowing down the jugular veins. Let us consider the mean blood pressure and flow rate in the arteries, microcirculation, and veins of the neck. For the purposes of this discussion, we will ignore the pulsatile nature of both pressure and flow rate and assume steady flow. In a rigid tube, neglecting inertia, the pressure difference between the upstream and downstream ends will be approximately related to the flow rate by the equation

$$P_1 - P_2 = RLQ + \rho gh, \tag{3.6}$$

where R is the (viscous) resistance per unit length, L is the length of the tube, Q is the flow rate, ρ is the fluid density, g is the acceleration due to gravity, and h is the vertical distance (positive if measured upwards) between the ends. Equation (3.6) is essentially Poiseuille's law with the inclusion of the gravitational term. As we have seen, the arteries will remain distended all the way up to the head and therefore will be effectively rigid, so it is reasonable to propose a simplified model of the system as sketched in Figure 3.9, which is also a sketch of an experiment that has been performed in the laboratory, with water in place of blood and a thin-walled flexible tube in place of the jugular vein (Hicks and Badeer 1989). In this model, P_1 is the pressure generated by the pump (the heart), and the wider vertical and curved tube represents the distended (and effectively rigid) carotid arteries and the vessels within the skull; it has length L and emerges from the skull at the point where the pressure is P_2, at a height h above the pump. Equation (3.6) applies.

If there were no tube downstream of the 'skull', the fluid would emerge into the atmosphere, so the pressure P_2 would be atmospheric; Equation (3.6) shows that there could be no flow unless $P_1 - P_{atm}$ were greater than ρgh. In other words, the fluid has to be pumped uphill and the flow rate (Q) would be

$$Q = (P_1 - P_{atm} - \rho gh)/RL. \tag{3.7}$$

On the downflow side, the fluid would merely fall under gravity; in their experiment, Hicks and Badeer (1989) collected the fluid from their flexible tube in a chamber at the same level as the pump. They observed that the tube collapsed uniformly to a very flattened shape with very small

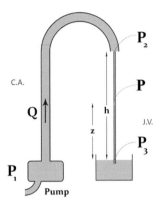

FIGURE 3.9 Sketch of an idealised model of the circulation in the neck of the giraffe, which also corresponds to the laboratory experiment performed by Hicks and Badeer (1989). The wider vertical tube (C.A.) and the curved tube represent the carotid artery and the vessels within the skull, which are taken to be distended and therefore effectively rigid. The narrow vertical tube (J.V.) represents the jugular vein, in which the pressure is close to atmospheric, and which is therefore highly collapsed – see the text for details. Thanks to Jon Pedley for the artwork.

cross-sectional area and the pressure within it remained uniform, equal to atmospheric pressure all the way along. Consequently $P_2 = P_3 = P_{atm}$ and the flow rate is given by Equation (3.7). Application of Equation (3.5) to the collapsed tube gives

$$P_2-P_3 = 0 = R_V hQ-\rho gh. \tag{3.8}$$

Here R_V is the mean resistance per unit length in the collapsible tube, and the sign of the gravity term is now negative, because h is measured downwards; note that the flow rate Q must be the same as in the 'artery' (what goes into the head must come out). It follows from Equation (3.8) that the resistance of the downflow tube must be given by $R_V Q = \rho g$; since Q is known, this determines that R_V must be large enough to balance the driving force of gravity. The resistance of a tube depends strongly on its cross-sectional area (as in Poiseuille's law), so this equation therefore determines the cross-sectional area of the collapsed tube.

In giraffes (as in humans) veins have much thinner walls than arteries, and the external pressure acting on a jugular vein is close to atmospheric, so the thin-walled flexible tube discussed above is a natural model to choose. We therefore expect the internal pressure in the jugular vein of the giraffe to remain close to atmospheric everywhere and the vein to be quite severely collapsed when the neck is upright. It cannot be the case that the arterial pressure in the skull, and at the exit from it, is significantly sub-atmospheric, since it would immediately collapse on exit and cause the internal pressure to become effectively atmospheric. The flow rate in the jugular vein, Q, will be determined by the pressure generated by the heart and by the resistances of the carotid arteries and the smaller vessels within the skull. Downstream, the system behaves more like a waterfall than a siphon. To summarise the contents of this section:

> The giraffe has a neck of phenomenal length
> Up which blood must be pumped to the brain;
> The consequence is a heart of great strength
> And collapse of the jugular vein

(T.J.P. 1996)

One final, intriguing, observation about giraffes is that some (but apparently not all) have valves in the jugular veins. These can hardly be present in order to encourage venous blood flow against gravity because it is gravity that drives the flow, so why are they there? The answer has to be to assist venous return when the animal puts its head down to drink, although if this assistance were really important, the valves should be present in all giraffes. This remains a puzzle. One may additionally note that, if a drinking giraffe has to raise its head quickly – if threatened by a predator for example – the pressure distribution would have to change rapidly, implying the existence of a sophisticated neural system to control the time-dependent dynamics. Investigating this may be an interesting research project for the future. Certainly, the topic of time-dependent flow-structure interactions in collapsible tubes is a fascinating area of fluid dynamics (Pedley and Pihler-Puzovic 2015).

Concluding Remarks

The author hopes that this chapter will have given the reader some understanding of the fluid dynamics of blood flow in human beings and other mammals: of how the pulse wave propagates in arteries, of how the viscosity of blood combined with its inertia leads to complex flow patterns and to distributions of shear stress exerted on the artery walls that may have an important influence on the development of atherosclerosis, of how the fact that blood is not a continuous fluid but consists of a suspension of cells is of great importance in the microcirculation, and of how the flexibility of veins combines with gravity to cause collapse of the giraffe's jugular veins. In addition, however, he also hopes to have given an appreciation of the role that rather simple mathematical models can play in acquiring this understanding.

The modelling of pulse-wave propagation is a good example – neglect many of the features that we know are present, such as viscosity and the non-uniformity of real arteries, isolate the two physical principles of con-servation of mass and Newton's laws of motion, together with a description of the wall elasticity, and we can predict the speed of propagation of the pulse wave with considerable accuracy. After noting that this model does not account for the observed peaking of the pressure pulse, we extended the model slightly to enable us to analyse wave reflection at a bifurcation and found that the model did then account for the peaking of the pulse. Adding blood viscosity would have allowed other predictions which could be compared (successfully) with other observations. This perfectly illustrates the essential property of applied mathematics, to enable one to answer questions posed from outside mathematics, and thereby increase one's physical understanding of important real phenomena. The details can then be filled in using numerical simulation on a computer.

Acknowledgements

I am indebted to Professor David N. Ku, of the Georgia Institute of Technology, for pointing me to two videos that had been produced in his laboratory, one on blood flow in complex arterial geometry and one on self-excited oscillations in a collapsible, stenosed tube, and that I was able to include in my Darwin Lecture.[3] The related work is published in Moore et al. (1992). I am also very grateful to Professor David Elad, from Tel Aviv University, for telling me about the pioneering science of Moses Maimonides in the twelfth century – see Rosen (1995).

---

[3] The videos can be viewed at timings about 48 and 101 minutes within the lecture at www.youtube.com/watch?v=7ZL9G6Z7-qQ.

References

Caro, C. G., Fitz-Gerald, J. M., and Schroter, R. C. (1971) 'Atheroma and arterial wall shear: observation, correlation and proposal of a shear-dependent mass transfer mechanism for atherogenesis'. *Proceedings of the Royal Society of London B* 177, 109–159.

Fry, D. L. (1968) 'Acute vascular endothelial changes associated with increased blood velocity gradients'. *Circulation Research* 22, 165–197.

Hagenbach, E. (1860) 'Über die Bestimmung der Zähigkeit einer Flüssigkeit durch den Ausfluß aus Röhren'. *Poggendorf's Annalen der Physik und Chemie* 108, 385–426.

Hargens, A. R., Millard, R. W., Pettersson, K., and Johansen, K. (1987) 'Gravitational haemodynamics and oedema prevention in the giraffe'. *Nature* 329, 59–60.

Harvey, W. (1628) *Movement of the Heart and Blood Vessels in Animals*, translated by K. J. Franklin (1957). Oxford: Blackwell.

Hicks, J. W., and Badeer, J. S. (1989) 'Siphon mechanism in collapsible tubes: application to circulation of the giraffe head'. *American Journal of Physiology* 256, R567–R571.

Ku, D. N. (1997) 'Blood flow in arteries'. *Annual Review of Fluid Mechanics* 29, 399–434.

McDonald, D. A. (1974) *Blood Flow in Arteries*, 2nd edition. London: Edward Arnold.

Moore, J. E. Jr., Ku, D. N., Zarins, C. K., and Glagov, S. (1992) 'Pulsatile flow visualization in the abdominal aorta under differing physiologic conditions: implications for increased susceptibility to atherosclerosis'. *American Society of Mechanical Engineers Journal of Biomechanical Engineering* 114, 391–397.

Olson, R. M. (1968) 'Aortic blood pressure and velocity as a function of time and position'. *Journal of Applied Physiology* 24, 563–569.

Pedley, T. J. (2003) 'Mathematical modelling of arterial fluid dynamics'. *Journal of Engineering Mathematics* 47, 419–444.

Pedley, T. J. and Pihler-Puzovic,D. (2015) 'Flow and oscillations in collapsible tubes: physiological applications and low-dimensional models'. In V. Shankar and V. Kumaran (eds.) *Proceedings of IUTAM Symposium Transition and Turbulence in Flow through Deformable Tubes and Channels*. Special issue, *Sadhana* 40(3), 891–909.

Rosen, S. D. (1995) 'Moses Maimonides, prince of physicians'. *Journal of the Royal College of Physicians* 29, 561–563.

Schroter, R. S., and Sudlow, M. F. (1969) 'Flow patterns in models of the human bronchial airways'. *Respiration Physiology* 7, 341–355.

Skalak, R., and Brånemark, P. I. (1969) 'Deformation of red blood cells in capillaries'. *Science* 164, 717–719.

West, J. B. (2008) 'Ibn al-Nafis, the pulmonary circulation, and the Islamic Golden Age'. *Journal of Applied Physiology* 105, 1877–1880.

Young, T. (1809) 'On the functions of the heart and arteries'. *Philosophical Transactions of the Royal Society of London* 99, 1–31.

Further Reading

Caro, C. G., Pedley, T. J., Schroter, R. C., and Seed, W. A. (1978) *The Mechanics of the Circulation.* Oxford: Oxford University Press (2nd edition, Cambridge: Cambridge University Press, 2011).

Fry, D. L. (1987) 'Mass transport, atherogenesis and risk'. *Arteriosclerosis* 7, 88–100.

Parker, K. H. (2009) 'A brief history of arterial wave mechanics'. *Medical and Biological Engineering and Computing* 47, 111–118.

Pedley, T. J. (1995) 'High Reynolds number flow in tubes of complex geometry with application to wall shear stress in arteries'. In C. P. Ellington and T. J. Pedley (eds.) *Society for Experimental Biology Symposium 49 Biological Fluid Dynamics.* Cambridge: The Company of Biologists, pp. 219–241.

Pedley, T. J., Brook, B. S., and Seymour, R. S. (1996) 'Blood pressure and flow rate in the giraffe jugular vein'. *Philosophical Transactions of the Royal Society of London Series B* 351, 855–866.

Vogel, S. (1983) *Vital Circuits.* Oxford: Oxford University Press.

4 *Dracula*, Blood, and the New Woman: Stoker's Reflections on the Zeitgeist

CAROL SENF

Students of the Long Nineteenth Century know that people during that time felt particularly ill equipped to handle the rapid changes taking place and that many felt like the persona of Matthew Arnold's poem *Dover Beach*. Speaking to his beloved, he thinks of sadness and human misery, concluding there is nothing for them but love because the larger world has 'neither joy, nor love nor light / Nor certitude, nor peace, nor help for pain' amidst the 'confused alarms of struggle and flight / Where ignorant armies clash by night.' (Arnold 1867). We, in 2021, may be less apprehensive about the speed with which change occurs, but we have our own angst to confront and can engage with what troubled or delighted our ancestors in the popular literature that most transparently reveals those concerns.

While almost everyone who writes on Stoker classifies him as a popular writer, or at least as the author of one enormously popular book, Glover puts Stoker's desire for popularity at the center of *Vampires, Mummies, and Liberals: Bram Stoker and the Politics of Popular Fiction* and observes that he must be read 'as he most wanted to be read, as a popular writer' (Glover 1996, 6). Although the current study focuses on *Dracula*, I share Glover's concern that readers overemphasise *Dracula* without recognising that Stoker's entire oeuvre addresses 'questions of nationhood, character, and sexuality, and the close links between them' (Glover 1996, 8), and I have elsewhere written on Stoker's interest in what troubled and delighted his contemporaries, including the rise of science, medicine, and technology, England's relationship to her colonies and fear of an uprising, changing gender roles, and the rise of the United States as a world power. The following chapter looks at a subset of these issues, focusing on the significance of blood and the impact of the New Woman.

While I am swayed by Glover's argument that Stoker experimented with popular forms to address social anxieties, his interest in contemporary issues was firmly ingrained, the result of family life and education. Although Bram followed his father (after whom he was named) into the Irish Civil Service after graduating from Trinity, his mother was the more influential parent. Charlotte Stoker was an incredibly strong woman, with ties to the Irish military, but she was also interested in improving the lives of the poor, especially the lives of poor women.

According to Peter Haining and Peter Tremayne, Charlotte spoke both for women and for the disabled, asked 'for equality of the sexes and the provision of dignified work for women', and wrote a pamphlet *On Female Emigration from Workhouses* (1864) that argued 'A self-supporting woman is alike respected and respectable' (Haining and Tremayne 1997, 46). She was a member of the Statistical and Social Inquiry Society of Ireland, and also published *On the Necessity of a State Provision for the Education of the Deaf and Dumb of Ireland* in 1863.

Influential in other ways, Charlotte was possibly the source of Bram's enthusiasm for horror. Being an invalid for the first seven years of his life, Bram had time to mull over her stories as well as those of his nurse. Haining and Tremayne cite Enid Stoker, the wife of Bram's brother Tom, who remembered Charlotte's telling of the 1832 cholera epidemic in Sligo, which Bram incorporates into 'The Invisible Giant' in *Under the Sunset* (Stoker 1882). Fearing both cholera and marauders, the family locked themselves in their home. When a looter attempted to break in, the intrepid Charlotte cuts off his hand with an axe. She also supported Bram's work, commenting enthusiastically that *Dracula* 'was splendid/a thousand miles beyond anything you have written before' (Murray 2004, 204). Responding to a positive review a month later, Charlotte reinforced her previous praise, saying that 'no book since Mrs Shelley's *Frankenstein* or indeed any other at all has come near yours in originality, or terror ... it should make a wide spread reputation and much money for you' (Murray 2004, 204).

Charlotte might have provided him with the first model of the New Woman (though that phrase would not become part of the lexicon for 20 years), and he worked with other independent women at the Lyceum. Trinity also encouraged his enthusiasm for current events, providing opportunities to explore them as a member of its most prestigious debating clubs, the Philosophical Society and the Historical Society. Elected President of the Philosophical Society in 1870 and Auditor (the highest office of the Historical Society) in 1872, Stoker was responsible for the accounts of the Historical Society as well as for delivering an address, 'The Necessity for Political Honesty', at the Philosophical Society's annual meeting.

After Trinity and work at Dublin Castle, Stoker moved to London, from civil service to theatre, from theatre reviews to writing popular fiction, which he did mostly during vacations until Irving died in 1905. *Dracula*, the focus of this lecture, is his most popular book as well as the work on which he spent the most time, as is evident in *Notes on Dracula* and other studies of its creation. *Dracula* has been explored with every conceivable theoretical lens, and critics have left no headstone unturned, but this lecture will focus on blood and the New Woman, blood revealing the degree to which Stoker was influenced by traditional forces, the latter pointing to the future.

Blood and *Dracula*

The single trait that connects all vampires until relatively recently is their need for blood as a source of sustenance, and Stoker's vampires are no exception, though they are much more in terms of metaphoric resonance. It is evident that Dracula goes hunting for something to feed his brides and equally apparent to Jonathan Harker that Dracula becomes rejuvenated by his diet:

> There lay the Count, but looking as if his youth had been half renewed, for the white hair and moustache were changed to dark iron-grey; the cheeks were fuller, and the white skin seemed ruby-red underneath; the mouth was redder than ever, for on the lips were gouts of fresh blood, which trickled from the corners of the mouth and ran over the chin and neck ... It seemed as if the whole awful creature were simply gorged with blood ... The coming night might see my own body a banquet in a similar way to those horrid three ... This was the being I was helping to transfer to London, where, perhaps, for centuries to come he might, amongst its teeming millions, satiate his lust for blood, and create a new and ever-widening circle of semi-demons to batten on the helpless.
>
> (Stoker 1983, 51)

Terrified that the brides will suck his blood, Jonathan also fears that he is responsible for transporting Dracula to England, where the vampire will have an endless supply of food. And Van Helsing later reminds his young followers that blood is what enabled Dracula to live for centuries.

Stoker, however, takes the fact that vampires across the globe are bloodsuckers and uses blood to reveal that vampires are creatures of the past. Not only is blood food for vampires, it signifies life itself, as Renfield observes to Dr. Seward, 'The blood is the life!' (Stoker 1983, 141). The loss of blood eventually results in death, as in the case of Lucy Westenra, but sharing blood with a vampire is also a kind of unholy communion that can also produce spiritual death for those who succumb.

Most importantly, though, in terms of Stoker's mulling over the events of his day, he uses blood to signify traditional family connections and thus to nod to the past. Shortly after arriving at Dracula's ruined castle, Harker is impressed by his host's stories about his noble ancestry: 'Why, there is hardly a foot of soil in all this region that has not been enriched by the blood of men, patriots or invaders' (Stoker 1983, 21). Indeed, Dracula is proud because his blood connects him to an ancient past:

> 'We Szekelys have a right to be proud, for in our veins flows the blood of many brave races who fought as the lion fights, for lordship . . . What devil or what witch was ever so great as Attila, whose blood is in these veins?' He held up his arms. 'Is it a wonder that we were a conquering race; that we were proud; that when the Magyar, the Lombard, the Avar, the Bulgar, or the Turk poured his thousands on our frontiers, we drove them back? . . . we of the Dracula blood were amongst their leaders . . . Ah, young sir, the Szekelys – and the Dracula as their heart's blood, their brains, and their swords – can boast a record that mushroom growths like the Hapsburgs and the Romanoffs can never reach. The warlike days are over. Blood is too precious a thing in these days of dishonourable peace; and the glories of the great races are as a tale that is told.
>
> (Stoker 1983, 29, 30)

Claudio Vescia Zanini devotes an entire critical work to blood in *Images of Blood in Bram Stoker's Dracula* that analyses the multiple meanings of blood: 'as food for the vampire, as a symbol of nobility or family connections, and as a means for blood bonds other than kinship' (Zanini 2013, 11).

Stoker, who, according to Murray, could trace his own lineage on his mother's side (the Blakes) back to the fourteenth century, nonetheless uses his portrayal both of Arthur Holmwood, his only English aristocrat, and of Dracula to suggest to readers that the aristocracy is trapped in the past. Arthur, for example, visits his father when the woman he loves is also very ill. Dracula, despite his vast library of contemporary books and newspapers, seems incapable of adapting to the world of modern steam transportation. The two are also linked by their use of traditional titles and power relationships, and Zanini reinforces the connection by observing that the literary vampires that might have influenced Stoker were aristocrats.

As Zanini notes, blood also connects the human characters, though often in ways that suggest that they are modern. Van Helsing, for example, comments on Seward's sucking the gangrenous blood from his wound, an act that links mentor and student and furthermore demonstrates their scientific knowledge. In addition, even though Stoker was not aware of blood typing, and though the transfusions of blood into Lucy do not save her life, these transfusions hint at the science of the day. For example, Van Helsing comments on the quality of Arthur's blood, revealing his awareness of contemporary science: 'He is so young and strong and of blood so pure that we need not defibrinate it' (Stoker 1983, 122). Kim Pelis explores the history of transfusion, which was often used to save the lives of women in danger of dying in childbirth, in nineteenth-century England, revealing information that Stoker might have learned from his physician brothers:

> For nearly 75 years, English transfusion practitioners had attempted to control blood by defibrinating it, diluting it, and modifying its instrumental route between bodies. They went to this trouble because ... they believed that blood alone could reanimate patients near death after blood loss. At the beginning and near the end of this period of blood-faith, English obstetricians turned to physiology for guidance about the fitness and methods of the operation. In the end, they found they could tame blood only by denying its vitality and, therefore, its therapeutic value. Small quantities of transfused blood could be replaced by far larger quantities of infused saline, turning what had been a grand and risky operation performed by highly trained experts into a relatively safe

procedure that could be easily mastered. During the 1890s, blood transfusion quietly dropped from British medical literature.

(Pelis 2001, 195)

While the transfusions fail to save Lucy, Stoker uses them to provide an erotic connection between Lucy and the men who love her and want to protect her:

> 'Just so. Said he not that the transfusion of his blood to her veins had made her truly his bride?'
>
> 'Yes, and it was a sweet and comforting idea for him.'
>
> 'Quite so. But there was a difficulty, friend John ... Then this so sweet maid is a polyandrist, and ... I, who am faithful husband ... am bigamist.'
>
> (Stoker 1983, 176)

Dracula too recognises the erotic connection of blood during the scene in which he forces Mina to drink, from his breast, blood that he had ingested from Lucy and which combines the blood of the men who had loved her: 'And you, their best beloved one, are now to me, flesh of my flesh; blood of my blood; kin of my kin; my bountiful wine-press for a while; and shall be later on my companion and my helper' (Stoker 1983, 288). As a result, Quincey Harker, the son of Mina and Jonathan, shares the blood of Dracula as well as the blood of the men who had joined to battle Dracula. His existence as well as his name suggests a connection in which the opponents of Dracula join in a new kind of family.

Stoker's multiple references to blood suggest other things as well. The vampires' diet of blood might remind readers of an unholy communion which leads to eternal death instead of eternal life, and numerous passages in the novel reinforce that blood is a reminder of human mortality. Early in the novel, Jonathan Harker is confronted by the three women in Dracula's castle:

> The fair girl advanced and bent over me till I could feel the movement of her breath upon me. Sweet it was in one sense, honey-sweet, and sent the same tingling through the nerves as her voice, but with a bitter underlying the sweet, a bitter offensiveness, as one smells in blood.
>
> (Stoker 1983, 38)

The reminder of mortality is even stronger when Harker describes Carfax Abbey, one of Dracula's English homes:

> There was an earthy smell, as of some dry miasma, which came through the fouler air. But as to the odour itself, how shall I describe it? It was not alone that it was composed of all the ills of mortality and with the pungent, acrid smell of blood, but it seemed as though corruption had become itself corrupt. Faugh! it sickens me to think of it. Every breath exhaled by that monster seemed to have clung to the place and intensified its loathsomeness.
>
> (Stoker 1983, 251)

Harker's reaction reminds readers that Dracula is entirely a creature of the body.

One more reference to blood in *Dracula* warrants our attention, the use of blood and/or bloody as a curse, in the novel used primarily by the working-class men encountered during their travels. I draw attention to it primarily because I don't think anyone has ever mentioned it before. Mina refers to Dracula's travels on the *Czarina Catherine*:

> No one knew where he went 'or bloomin' well cared', as they said, for they had something else to think of – well with blood again; for it soon became apparent to all that the *Czarina Catherine* would not sail as was expected . . . The captain swore polyglot – very polyglot – polyglot with bloom and blood; but he could do nothing . . . Then the captain replied that he wished that he and his box – old and with much bloom and blood – were in hell.
>
> (Stoker 1983, 318)

This scene is certainly not important, though it reveals that Stoker was a conscious craftsman who reminds readers of the centrality of blood as a reminder of the connection to the earth and the limitations of the human body. Vampires and working-class men are also characterised by their thirst, but that's a connection for another study.

The New Woman in *Dracula*

If Stoker's treatment of blood looks backward at history and tradition, the New Woman is *au courant*, a modern woman who is critical of tradition, especially the belief that women should be subordinate to men. Most proponents of the New Woman believed that men and women should be equal in all things and, therefore, that women should have access to education and the professions. The one area in which some New Women disagreed was whether women should be as free as men to openly express their sexuality. Sarah Grand (1894), who coined the phrase in a debate with fellow writer Ouida, and whose character Angelica proposes to her future husband in *The Heavenly Twins* (1893), urges chastity for both men and women, and *The Heavenly Twins* features one negative consequence of sexuality when Edith contracts syphilis from her licentious husband and dies after giving birth to an infected child. It's possible to read vampirism as a similar kind of contagious disease, but Stoker echoes Grand's marriage proposal both in *Dracula* and in *The Man*. That Stoker is critical of these proposals might cause readers to assume he was hostile to the New Woman, and I confess I initially shared that prospective. Glover, for example, describes both *The Man* and *Lady Athlyne* as 'angry ripostes to the growing movement for women's suffrage' (Glover 1996, 20). Richard Dalby and William Hughes refer to Stephen Norman's behaviour in *The Man* when she proposes to a man and wants to be her father's son as both fascinating and repulsive, and argue that by 1905 'the age of the New Woman was effectively over' (Dalby and Hughes 2004, 21), a point with which both Sally Ledger, a leading expert on the New Woman, and I disagree.

If I once regarded Stoker as largely hostile to the New Woman, today I find his response to be more nuanced. Not only was Stoker close to his mother until she died, he was also friends with a number of other independent professional women, including the actresses Ellen Terry and Genevieve Ward and New Women writers George Egerton (whose name he borrows for a character in *The Man*), Elizabeth Robins, and Grant Allen.

Two excellent resources on the New Woman are Ledger and Gillian Sutherland, though the two emphasise different aspects, with Ledger focusing on fictional treatments and Sutherland exploring employment and education. Ledger begins with a precise historical overview:

> The New Woman was not 'named' until 1894, in a pair of articles by Sarah Grand and Ouida ... Gender was an unstable category at the *fin de siècle*, and it was the force of gender as a site of conflict which drew such

> virulent attacks upon the figure of the New Woman. These attacks were, in turn, closely related to the high profile of the Victorian women's movement in the late nineteenth century.
>
> This study will focus on the 1880s and 1890s, when the New Woman had her heyday.
>
> (Ledger 1997, 2)

An earlier article by Ellen Jordan asserts that the New Woman was criticised 'for aping the customs and habits, and even rivalling the physical strength, of men' (Jordan 1983, 20) and that the comic publication *Punch* was largely responsible for this hostility. Sutherland focuses primarily on women's education and work, and she moves from the caricatured woman in bloomers to examine their economic independence:

> This study is therefore an attempt to re-situate the New Woman caricature and the broader debate of which it was a part, by examining the opportunities for earning money, achieving independence, available to middle-class women in Britain in the last third of the long nineteenth century, from about 1870 to the outbreak of the First World War.
>
> (Sutherland 2015, 8)

Stoker, familiar both with his mother's insistence that women are able to support themselves and with professional women writers and actresses, creates in Mina Harker a woman character who is financially independent before her marriage.

That Stoker is thinking about the significance of the New Woman is evident early in *Dracula* when Mina reflects on this cultural phenomenon. In Whitby on vacation with Lucy, Mina describes a long walk followed by 'a capital "severe tea" at Robin Hood's Bay' and notes that their appetites 'should have shocked the "New Woman"', an observation that suggests their behaviour is not exactly ladylike (Stoker 1983, 88). She follows this observation of unladylike behaviour by echoing *The Heavenly Twins*, in which Angelica Hamilton Wells proposes to her future husband:

> If Mr Holmwood fell in love with her seeing her only in the drawing-room, I wonder what he would say if he saw her now. Some of the 'New Woman' writers will some day start an idea that men and women should be allowed to see each other asleep before proposing or accepting. But

> I suppose the New Woman won't condescend in future to accept; she will
> do the proposing herself. And a nice job she will make of it too! There's
> some consolation in that.
>
> (Stoker 1983, 88–89)

While Mina works as an assistant school mistress, is learning shorthand to
help her husband in his job, is an accomplished typist, and resents men who
try to protect her, Stoker reinforces that she is far from the sexually
aggressive or mannish New Woman mocked in *Punch*. Even after her
marriage, she is concerned at being seen with a man holding her by the arm:

> I felt it very improper, for you can't go on for some years teaching
> etiquette and decorum to other girls without the pedantry of it biting into
> yourself a bit; but it was Jonathan, and he was my husband, and we didn't
> know anybody who saw us – and we didn't care if they did – so on
> we walked.
>
> (Stoker 1983, 171)

Mina's concern over impropriety here echoes her concern at finding the
barefoot Lucy sleepwalking around Whitby in her nightgown, an episode
on which Ledger comments: 'Mina "reflects in her journal that 'I was filled
with anxiety about Lucy, not only for her health, lest she should suffer
from exposure, but for her reputation in case the story should get wind'"'
(Ledger 1997, 101). Ledger goes on to explain Mina's concern by adding:
'Unescorted women in the nineteenth century . . . were regularly mistaken
for prostitutes: "nice" women never went out unaccompanied, and this
explains Mina Murray's concern for her friend' (Ledger 1997, 101).

Despite her apparent criticism of the New Woman, Mina shares many
of the characteristics associated with her. Not only is she employed, but
she is also learning the clerical skills often associated with the New
Woman, though she is careful to note that she plans to use them to assist
Jonathan in his work:

> When we are married I shall be able to be useful to Jonathan, and if I can
> stenograph well enough I can take down what he wants to say in this way
> and write it out for him on the typewriter, at which I am also practicing
> very hard.
>
> (Stoker 1983, 53)

Mina goes on to say that she had been influenced by journalists (many of whom would have been New Women):

> I shall try to do what I see lady journalists do: interviewing and writing descriptions and trying to remember conversations. I am told that, with a little practice, one can remember all that goes on or that one hears during a day.
>
> (Stoker 1983, 54)

Indeed, Mina uses the technical and organisational skills that many readers would have associated with the New Woman to assemble everything Dracula's opponents have learned about him to track him down and destroy him.

Even though Mina apparently plans to be a traditional support to her husband, her belief in her equality is also evident when she bridles when the men treat her as an inferior when they decide to protect her from knowing too much about their quest to track Dracula to his home.

> All the men, even Jonathan, seemed relieved; but it did not seem to me good that they should brave danger and, perhaps, lessen their safety – strength being the best safety – through care of me; but their minds were made up, and, though it was a bitter pill for me to swallow, I could say nothing, save to accept their chivalrous care of me.
>
> (Stoker 1983, 242)

Certainly, later events prove that the desire to protect her is misguided. Left alone in Dr Seward's hospital, Mina is attacked by Dracula. Unlike her friend Lucy, however, Mina consciously fights against Dracula's power over her.

While Mina shares a number of characteristics with the New Woman, her friend Lucy, who succumbs to Dracula's advances while sleepwalking in Whitby and becomes his first English victim, is presented as having only one characteristic of the New Woman, that of being sexually adventurous. Telling Mina about receiving three proposals in one day, she confesses a desire that women be allowed to have multiple husbands, though she quickly stops herself by noting that 'this is heresy, and I must not say it' (Stoker 1983, 59). Nonetheless, this desire for three husbands is Stoker's way of connecting her to what Ledger describes as 'the sexually decadent New Woman' like 'the three female vampires at Castle Dracula' (Ledger 1997, 101), though she struggles against those sexual urges while she is conscious, succumbing to them only as she comes more and more under Dracula's control. Towards the end of her life, she is as fully seductive as Dracula's other brides, as Seward notes to his horror:

> And then insensibly there came the strange change which I had noticed in the night. Her breathing grew stertorous, the mouth opened, and the pale gums, drawn back, made the teeth look longer and sharper than ever. In a sort of sleep-waking, vague, unconscious way she opened her eyes which were now dull and hard at once, and said in a soft voluptuous voice . . .
> 'Arthur! Oh, my love, I am so glad you have come! Kiss me!'
>
> (Stoker 1983, 161)

Fortunately, Van Helsing is there to protect Arthur from acceding to his desire for a final kiss. Despite Van Helsing's best efforts to keep her from becoming a vampire, however, Lucy enters the realm of the undead, at which time she manifests yet another characteristic attributed to the New Woman, her lack of motherliness.

Like the women in Dracula's castle who receive a bag from which Harker hears what he surmises is 'a gasp and a low wail, as of a half-smothered child' (Stoker 1983, 39), Lucy begins her vampiric career by feeding on children who refer to her as the 'bloofer lady'. Ledger comments on the fact that 'the New Woman was often figured in discourse as at best a bad mother and at worst an infanticidal one' (Ledger 1997, 104). Dr Seward, who sees her with a child victim, is horrified by her reaction: 'With a careless motion, she flung to the ground . . . the child that up to now she had clutched strenuously to her breast, growling over it as a dog growls over a bone' (Stoker 1983, 211).

The same scene reveals that Lucy is more than happy to graduate from children to adult prey, which convinces Dr Seward that the woman he had once loved and wanted to marry is a monster who must be killed: 'At that moment the remnant of my love passed into hate and loathing; had she then to be killed, I could have done it with savage delight' (Stoker 1983, 211). Indeed, Lucy and the three women in Dracula's castle are all destroyed in bloody scenes that suggest a degree of total hostility to the sexually aggressive New Woman.

What makes Stoker's handling of the New Woman so interesting is his treatment of Mina. Not only does she accompany the men to Castle Dracula, but she is regarded as an equal during much of the quest to destroy Dracula. She also witnesses Dracula's destruction, noting the possibility of redemption: 'It was like a miracle; but before our very eyes, and almost in the drawing of a breath, the whole body crumbled into dust and passed from our sight' (Stoker 1983, 377). Dracula's destruction and the removal of the scar from Mina's forehead would have been a perfectly sensible conclusion. Stoker, however, returns to more traditional ideas of gender when he has the group return to the same location seven years later. Harker's brief note remarks that Arthur and Seward are happily married and that he and Mina have a son whose 'bundle of names links all our little band of men together' (Stoker 1983, 378). Most problematic, however, is that the note effectually reduces Mina to the traditional roles of wife and mother. Having collected the material that enabled them to understand Dracula's movements and suffered for the men's mistaken ideas of chivalry, she is effectively silenced and returned to a traditional role for women. On the other hand, what reader of *Dracula* can ever forget her strengths? Instead of being the typical Gothic heroine whose role in the novel is only to be rescued, Mina demonstrates new possibilities for women, if only for a little while.

The New Woman in Stoker's Later Fiction

Stoker's treatment of the women in *Dracula* is a disappointment for many modern readers, since he kills off four of his women characters and hobbles the fiercely independent Mina Harker at the conclusion. His future books demonstrate that Stoker was not finished with thinking about the New Woman. Indeed, he continues to focus on New Women characters in *The Man* (1905) and *Lady Athlyne* (1908). Enthusiastically reviewed in *Punch* and *The Bookman*, *The Man* has attracted very little attention since, the exceptions being Roth, whose critical study of Stoker characterises it as 'the most direct and lengthy analysis of female sexuality and behavior Stoker undertook' (Roth 1982, 38), Glover, and Radclyffe Hall's *The Well of Loneliness* (1928), which echoes early sections of *The Man*. Stoker's heroine is Stephen Norman, and Hall's female protagonist Stephen Gordon, and both Stephens try to be their fathers' sons, question conventional gender roles, and inherit property when their fathers die. Stoker's heroine is decidedly heterosexual, however, in contrast to Hall's lesbian protagonist.

An only child whose mother died at her birth, Stephen attempts to follow in her father's footsteps but often encounters problems because of her gender. Wishing to know more about her tenants by accompanying her father to the Petty Sessions Court, she is told by her very traditional aunt that it would be improper for a young woman to hear 'low people speaking of low crimes . . . cases of low immorality; cases of a kind . . . that you are not supposed to know anything about' (Stoker 1905, 46). Visiting an old friend of her mother, Mrs Egerton (Stoker was friends with the New Woman writer Mary Chavelita Dunne Bright, who wrote using the pen name George Egerton), a professor at Somerville in Oxford, reinforces the traditional gender roles that emphasised women's inferiority:

> If, indeed, she was a woman, and had to abide by the exigencies of her own sex, she would at least not be ruled and limited by woman's weakness. She would plan and act and manage things for herself, in her own way . . .
> Fortunately her father . . . wished her to grow up in manly ways. At last she seemed to understand something of his purpose in her peculiar education. How strong and good it was! How she would henceforth bend herself to it; and would so fashion her own acts that they would not be tinged with the woman's weakness of which she had had so painful an experience!
> Whatever her thoughts might be, she could at least control her acts. And those acts should be based not on woman's weakness, but on man's strength!
>
> (Stoker 1905, 43)

Believing in women's equality and learning that her spinster aunt had turned down several proposals but remained unmarried because the man she loved most did not propose, and women of her generation were expected to remain silent, Stephen concludes that women's unhappiness stems from their passivity:

> Why should good women's lives be wrecked for a convention? Why in the blind following of some society fetish should life lose its charm, its possibilities? Why should love eat its heart out, in vain? The time will come when women will not be afraid to speak to men, as they should

> speak, as free and equal. Surely if a woman is to be the equal and lifelong companion of a man ... she should be free at the very outset to show her inclination to him just as he would to her.
>
> (Stoker 1905, 53–54)

Determined not to be passive, Stephen thus decides on a bold experiment, to propose to an old friend, Leonard Everard. However, unlike Grand's Angelica, who proposes to a man who loves and supports her, Stephen's bad choice turns her down and then, having reconsidered, attempts to blackmail her because he needs her money to pay his debt, a plan that Stephen and her aunt manage to circumvent.

The remainder of the novel involves Stephen's realisation that a better choice is Harold An Wolf, the Man of the title, and *The Man* concludes in typical romance fashion with Harold and Stephen declaring their love for one another: 'In that divine moment, when their mouths met, both knew that their souls were one' (Stoker 1905, 332). While this ending might suggest the triumph of traditional gender roles, *The Man* also implies that much has changed in the personal and social circumstances surrounding Stephen and that Stephen is a New Woman.

Stoker handles this transformation subtly in ways that echo his exploration in *Dracula*. The beginning of the novel emphasises the power of the past, including the importance of blood in both Stephen and Harold. Stoker emphasises Harold's Saxon ancestry, and Stephen's surname, Norman, points to the past, though her background is more complicated:

> The glorious mass of red hair, full, thick, massive, long and fine of the true flame colour, showed the blood of another ancient ancestor of Northern race ... The black eyes, deep blue-black, or rather purple-black, the raven eyebrows and eyelashes, and the fine curve of the nostrils spoke of the Eastern blood of the far-back wife of the Crusader.
>
> (Stoker 1905, 7)

Stoker emphasises that neither Stephen nor Harold is limited by their past. Harold, for example, goes to the gold fields of North America and acquires a fortune, while Stephen inherits the Earldom of Lannoy from a distant relative. While this inheritance might signify a link to the past, Stoker asks readers to see it as an indication of something new because

Stephen has a lifelong desire for justice, and she is not hampered by conventional gender roles. She inherits the earldom through the female line because the world is changing, as Stoker spells out in his discussion of the results of the Boer War:

> Towards Christmas in the second year the Boer war had reached its climax of evil ... In the early days the great losses were not amongst the troops but their leaders; the old British system of fighting when ... officers led on their men had not undergone the change necessitated by a war with a nation of sharpshooters.
>
> (Stoker 1905, 247)

Technological changes in warfare had destroyed many great English families, and Stephen reads in *The Times* that the Earl of Lannoy had died, having 'never recovered from the shock of hearing that his two sons and his nephew had been killed in an ambush by Cronje' (Stoker 1905, 248–249). Because there is no male heir, Stephen inherits Lannoy and consequently gains in rank (her father had been a squire), becoming a countess, the female equivalent of earl. The novel had opened with the adolescent Harold observing to Stephen that women were incapable of administering justice and with Stephen's musing, 'To be God and able to do things!' (Stoker 1905, 12). The novel recognises that, as a powerful landowner, she will be able to accomplish many things.

Lady Athlyne (1908) is the least studied of Stoker's novels, but it has the most to say about the New Woman and the equality that women were gaining as a result of the Boer War and the technology that levelled the playing field and created equal opportunities for men and women. Its heroine, Joy Ogilvie, comes from a very traditional American family. Her description of her father reveals his ties to the past:

> In Kentucky we still hold with the old laws of Honour which we sometimes hear are dead ... in other countries. My father has fought duels all his life. The Ogilvies have been fighters way back to the time of the settlement by Lord Baltimore.
>
> (Stoker 2007, 18–19)

She meets her future husband, the Earl of Athlyne, when he rescues her from a runaway horse. What's interesting, however, is that their future will revolve around the automobile instead of the horse, a technology with which both Joy and Athlyne are proficient, while her father is not. In fact, the automobile enables Joy and Athlyne to discover their love for one another away from the prying eyes of the older generation and to share that passion without shame. Such independence gives them freedom to express their feelings for one another openly. Unlike Mina and Jonathan Harker, who are sexually timid (Mina is even reluctant to be seen in public holding her husband's hand), Joy and Athlyne openly express their passionate feelings.

A typical romance, *Lady Athlyne* is full of misunderstandings among the generations, and the irascible Colonel Ogilvie even threatens Athlyne with a duel for damaging his daughter's reputation, until it is discovered that, due to a quirk of Scottish law, Joy and Athlyne are already married. Murray suggests that Stoker had been thinking about this law for some years.[1]

As an attorney, Stoker would have realised that the Scottish marriage law implies the equality of young people choosing one another freely rather than depending on families to choose life partners for them. It is inherently future-oriented.

Stoker emphasises other characteristics to signify that Joy and Athlyne represent the future. Unlike Arthur Holmwood, who is beholden to his father's wishes, Athlyne is an orphan whose experiences in the Boer War make him critical of tradition. Similarly, Joy brings with her a new kind of American spirit. Much as she respects her father's connection with tradition, she will make for herself a new way. What is most interesting, though, is that Joy, Stoker's last experiment with the New Woman character, combines all aspects of the character: a desire for equality and an enthusiasm for sexuality that doesn't undermine her desire for children.

---

[1] In a letter replying to Stoker's queries regarding the marriage laws in Scotland, the Bishop of Edinburgh informed the author that 'perfectly valid marriages' can be contracted by a couple's simply declaring themselves to be man and wife in front of witnesses and then ratifying the agreement before a magistrate. The Rt Rev. John Dowden to Bram Stoker, 20 January 1901 (Edinburgh), Brotherton Collection, Leeds University Library. Part of the importance of this letter is that it shows that Stoker was thinking about key episodes in *Lady Athlyne* at least seven and a half years before the novel was published (Murray 2004, 184).

A creation of the 1890s, the New Woman remained influential into the twentieth century as Ledger observes, noting 'the New Woman's continuing cultural significance in the early twentieth century' (Ledger 1997, 2). Ledger goes on to mention writers who explore the New Woman: Arnold Bennet, H. G. Wells, Joseph Conrad, D. H. Lawrence, and Virginia Woolf. By the early twentieth century, the figure that was often criticised and made the butt of jokes in the 1890s had become a model of womanly behaviour.

Though Stoker explores a number of popular genres and uses conventions from all of them, he is consistently forward-looking. His novels explore characters who are trapped by their past, often underlying their ties either to family tradition or to biology, the power of blood. In his most interesting novels, including those that feature the New Woman, he demonstrates an enthusiasm for new ideas and new projects, especially those that take advantage of new technology. Traversing the *fin-de-siècle*, his fiction demonstrates insights into a more enthusiastic perspective on the Zeitgeist at the end of his career than it did at the beginning.

References

Arnold, Matthew (1867) *New Poems.* London: Macmillan and Co., pp. 112–114.

Dalby, Richard, and Hughes, William (2004) *Bram Stoker: A Bibliography.* Southend-on-Sea: Desert Island Books.

Eighteen-Bisang, Robert, and Miller, Elizabeth (2008) *Bram Stoker's Notes for Dracula: A Facsimile Edition.* Jefferson, NC: McFarland.

Frayling, Christopher (2014) 'Mr. Stoker's Holiday'. In Jarlath Killeen (ed.) *Bram Stoker: Centenary Essays.* Dublin: Four Courts Press, pp. 179–200.

Glover, David (1996) *Vampires, Mummies and Liberals: Bram Stoker and the Politics of Popular Fiction.* Durham, NC: Duke University Press.

Grand, Sarah (1894) 'The New Aspect of the Woman Question'. *North American Review* 158, 270–276.

Haining, Peter, and Tremayne, Peter (1997) *The Un-dead: The Legend of Bram Stoker and Dracula.* London: Constable.

Jordan, Ellen (1983) 'The Christening of the New Woman: May 1894'. *The Victorian Newsletter* 63, 19–21.

Ledger, Sally (1997) *The New Woman: Fiction and Feminism at the fin de siècle*. New York: St Martin's Press.

Murray, Paul (2004) *From the Shadow of Dracula: A Life of Bram Stoker*. London: Jonathan Cape.

Pelis, Kim (2001) 'Blood Standards and Failed Fluids: Clinic, Lab, and Transfusion Solutions in London, 1868–1916'. *History of Science 39*, 185–213.

Roth, Phyllis A. (1982) *Bram Stoker*. Boston, MA: Twayne.

Stoker, Bram (1882) *Under the Sunset*. London: Sampson Low, Marston, Searle & Rivington.

Stoker, Bram (1983) *Dracula*. New York: Oxford University Press.

Stoker, Bram (1905) *The Man*. London: William Heinemann.

Stoker, Bram (2007) *Lady Athlyne*. Southend-on-Sea: Desert Island Books.

Sutherland, Gillian (2015) *In Search of the New Woman: Middle-Class Women and Work in Britain 1870–1914*. Cambridge: Cambridge University Press.

Zanini, Claudio Vescia (2013) *Images of Blood in Bram Stoker's Dracula*. New York: Lambert Academic Publishing.

5 Blood Lines of the British People

WALTER BODMER

Introduction – ABO and Other Blood Types

If you have ever been a blood donor you will probably know about ABO blood types and possibly know your own type. These are the inherited chemical differences on the surface of your red blood cells that need to be matched to avoid damaging reactions between your donor blood cells and the serum of the recipient. The A, B, and O blood types were first described by Landsteiner (1900), in the year of the rediscovery of Mendel's work on peas, which laid the foundations of our understanding of the mechanisms of inheritance. Landsteiner's experiment was very simple. He mixed the red blood cells of one person with the serum of another and did this pairwise for a number of people. To his surprise, he found that certain combinations of serum and red cells led to clumping or agglutination of the red cells. On the basis of resulting patterns of agglutination of the red cells, three types of serum were identified: A (which agglutinates only B cells), B (which agglutinates only A cells), and O (which agglutinates both A and B cells) as shown in Figure 5.1(a). These types were intrinsic to the individual and did not change with time or, for example, the presence of infections. They were, as we now know, one of the inherited characteristics of an individual's red blood cells. Landsteiner's work, together with many subsequent developments, meant that safer blood transfusions became possible.

It is surprising that Landsteiner did not immediately recognise that his A and B types were likely to be inherited as simple Mendelian dominant traits. This discovery was made by von Dungerne and Hirschfeld (1910) in a simple study of the inheritance of A and B in families, which showed that each was dominantly inherited, whereas O was recessive. For example, if a person who was AB, having inherited A from one parent and B from the other, married someone who was O, then on average half their children would be A and half B. This simple but fundamental analysis was the initial stimulus for all subsequent studies of the frequency of genetic variation in different human populations and the use of genetic variants for characterising genetic differences among humans.

The A and B types became the basis for the first study of genetic variation in human populations carried out by the Hirschfelds during the First World War (Hirschfeld and Hirschfeld 1919). They say that it was 'clear to us from the beginning that we could only attack the human race problem on serological lines' by using the 'iso-agglutinins first analyzed by Landsteiner'. By 'the human race problem' the Hirschfelds simply meant an understanding of the relationships between different human populations. Their results (Figure 5.1(b)), based on the assumption that A and B were inherited separately, showed substantial differences in the frequencies of A and B between different populations and major ethnic groups, with much higher frequencies of B in Indians than in the English. The Hirschfelds' frequencies were quite comparable to the frequencies found many years later using more data and better techniques. They concluded that there were two different races of humans, the As, who came from northern central Europe, and the Bs, who came from India, and that these two races mixed in the middle, namely between Europe and India. To us, this now seems a ludicrous conclusion and shows how mistaken one can be when placing too much emphasis on any single genetic polymorphism, a mistake that is still sometimes made.

It is clear that better genetic definition of human population variation required data on more than just one genetically variant system. The answer came from the discovery, especially in the 1940s and 1950s, of a whole range of new inherited blood group systems like ABO. Cavalli-Sforza and Edwards (1965) used the frequencies of these different blood types in a range of different human populations to characterise the genetic relationships between the populations. By using the similarities and differences between the blood group frequencies in these populations as a measure of genetic 'distance' between the populations, and some clever analysis, they were able to construct an evolutionary tree, the first ever of its sort, relating these populations (Figure 5.1(c)). This evolutionary, or

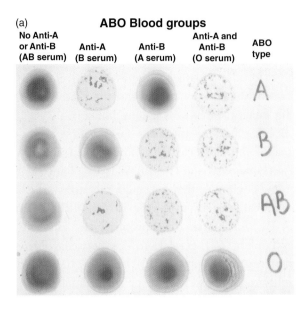

FIGURE 5.1 The discovery of the ABO and other blood groups and their use for population studies. (a) Landsteiner's experiment that identified the ABO blood groups. When the red blood cells of a person are mixed with the serum of another, certain combinations lead to the clumping of the red cells. A uniform red background in the circle, or well, indicates lack of clumping as the red cells are much too small to be seen individually. Clumping is shown by small patches of red in a well, which are aggregates of red cells clumped by an antibody. Thus, a person of type A has antibodies against B and so clumps the red cells of people who are either B or AB. Similarly, a person of type B has anti-A antibodies and clumps the red cells of people who are either A or AB. AB people have neither anti-A nor anti-B, while O individuals, who are neither A nor B, have antibodies of both types. On the basis of these patterns of clumping it is easy to identify the four ABO types, A, B, AB, and O. If, for example, group A red cells are infused into a recipient who is of group O, the recipient's anti-A antibodies bind to the transfused group A cells, leading to a potentially very severe reaction. That is why ABO matching is important for successful blood transfusion. It also turns out to be important for organ transplantation of, for example, kidneys or hearts. (b) The Hirschfelds' data on the worldwide distribution of the A and B blood types. The bar graph shows the frequencies of A and B obtain by the Hirschfelds in a wide range of populations of soldiers in the First World War. Note the gradient in the frequency of B, going from low in English and other Europeans to high in Indians (Hirschfeld and Hirschfeld 1919). (c) The first human evolutionary tree. This tree is based on gene frequency data for five blood group systems (Cavalli-Sforza and Edwards 1965).

(b)

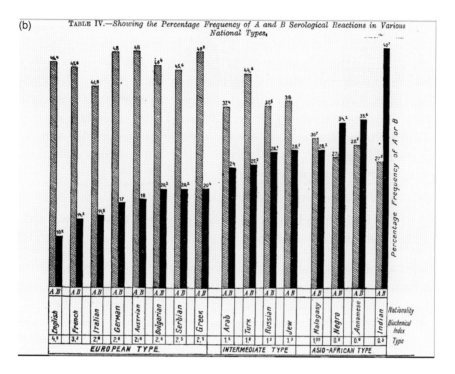

TABLE IV.—*Showing the Percentage Frequency of A and B Serological Reactions in Various National Types.*

(c)

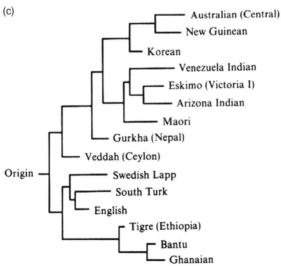

FIGURE 5.1 (*cont.*)

phylogenetic, tree was an outstandingly original piece of work that has formed the basis of all subsequent phylogenetic analyses based on genetic variation. The tree was obtained using data from only five blood groups and 18 types on 15 populations. In their words, it showed 'three branches, Europeans, Africans and Asiatics . . . splitting into a variety of subgroups' and a 'remarkable similarity between the blood group map thus obtained and the geographical map'. It was a surprising and remarkable achievement to obtain a result that was so consistent with general expectations using relatively few genetic variants.

The Human Leukocyte Antigen System Tissue Types and Disease Associations

Essentially the only graft transplants, for example, of skin, kidneys, or bone marrow, between any two individuals that survive without medication are those between identical twins. This indicates that there must be many inherited differences between any two individuals that the recipient's immune system recognises on the donor tissue as foreign. This observation, in turn, led to a search for blood-group-like systems on the white cells of the blood, on the assumption that such differences were likely to be present on most cells in the body and so might account for graft rejection. The result of the search was the discovery of the human leukocyte antigen (HLA) system, which is used to match donors and recipients to help avoid graft rejection and which plays a major role in the function of the immune system. The HLA system is the most variable genetic system in the human genome. It consists of six genes that sit more or less next to each other as a cluster on the sixth human chromosome. Each gene has a large number of variants, more than 100 in some cases (see Figure 5.2(a) for an early version of the information). Soon after the HLA system was discovered, it became a major additional source of genetic variability for the study of population relationships. The fact that the genes of the HLA system are close together on the same chromosome means that, despite the enormous variability, there are only four different type combinations in a family (Figure 5.2(a), lower right inset), which makes it possible to find pairs of siblings that are mutually fully compatible as donors and recipients for transplantation.

The close chromosomal juxtaposition of the HLA genes leads to population associations between the HLA types, as illustrated in

Figure 5.2(b). Thus, the UK population frequency of the variant *A1* of the *HLA-A* gene is 0.178 and the frequency of *B8* at the nearby *HLA-B* is 0.128. If they were independent of each other, the expected frequency of the combination *A1–B8* would be $0.178 \times 0.128 = 0.023$, whereas the observed frequency is 0.071, more than three times that expected assuming independence. This tendency for associations between the frequencies of genetic variants at closely juxtaposed genes is called linkage disequilibrium (LD), and it plays an important role in the analysis of genetic variation in populations as well as in the interpretation of associations between genetic variants and diseases as discussed below.

The role of the HLA system in immune responses suggested that HLA might be involved in inherited susceptibility to autoimmune diseases such as coeliac disease (gluten sensitivity) and rheumatoid arthritis. The simplest way to investigate this was to find out whether there was an HLA type that was much commoner among people with a particular disease than among those without the disease. Some results of such a study are shown in Figure 5.2(c). For example, the HLA type Dw3 was found in more than 90 per cent of people with coeliac disease but in only about 25 per cent of the controls without the disease. Most notably, HLA B27 was found in nearly all people with ankylosing spondylitis, a form of arthritis in which there is inflammation of the joints of the spine, often leading to their fusion and a severely crooked back. In contrast, the frequency in non-diseased controls is generally less than 10 per cent. Four of the diseases shown in Figure 5.2(c) clearly involve abnormal immune activity, and the HLA association indicates a significant genetic contribution to their occurrence. The fifth disease, haemochromatosis, is associated with a variant in a gene, *HFE*, close to the HLA genes, which leads to abnormal control of iron metabolism. *HFE* was found by looking for a variant in a gene close to *HLA-A* which had functional relevance to haemochromatosis, and which could explain the association with *HLA-A3* by LD, as explained in the example given in Figure 5.2(b).

Following the identification of DNA as the genetic material, and the analysis of its structure, it became possible to study genetic variation at the DNA level as variations in the DNA sequence, rather than studying the proteins that are derived from the DNA sequence. Given that there are about three billion letters of the four-letter alphabet of the DNA

(a)

HLA

A1	B5	Bw47	Cw1	DR1	DQw1	DPw1
A2	B7	Bw48	Cw2	DR2	DQw2	DPw2
A3	B8	B49	Cw3	DR3	DQw3	DPw3
A9	B12	Bw50	Cw4	DR4		DPw4
A10	B13	B51	Cw5	DR5		DPw5
A11	B14	Bw52	Cw6	DRw6		DPw6
Aw19	B15	Bw53	Cw7	DR7		
A23	B16	Bw54	Cw8	DRw8		
A24	B17	Bw55		DRw9		
A25	B18	Bw56		DRw10		
A26	B21	Bw57		DRw11		
A28	Bw22	Bw58		DRw12		
A29	B27	Bw59		DRw13		
A30	B35	Bw60		DRw14		
A31	B37	BW61		DRw52		
A32	B38	Bw62		DRw53		
Aw33	B39	Bw63				
Aw34	B40	Bw64				
Aw36	Bw41	Bw65				
Aw43	Bw42	Bw67				
Aw66	B44	Bw70				
Aw68	B45	Bw71				
Aw69	Bw46	Bw72				
		Bw73				
		Bw4				
		Bw6				

A31
B7
Cw7
DRw13
DQw1
DPw4

FIGURE 5.2 The HLA system. (a) There are six major genes, *A*, *B*, *C*, *DR*, *DQ*, and *DP*. Each gene has many variant types, defined by a letter followed by a number, for example, *A1*, and each type is inherited as a simple dominant, just like A or B of the ABO blood groups. The six genes are sufficiently close to each other on chromosome 6 that they are nearly always inherited together in a combination known as a 'haplotype'. Each parent has two haplotypes, as indicated by the schematic shown in the inset. In the bottom right inset, the two bars on the left, with the green and blue squares at the top, represent one parent, and the two bars on the right, with the yellow and red squares at the top, represent the other parent. Each child inherits one haplotype from the mother and one from the father. Thus, as indicated in the schematic, each child can have only one of four types of combinations, namely green with yellow, green with red, blue with yellow, and blue with red, each of which occurs with the same probability of ¼. That means that the chance of finding a complete match for an individual amongst their siblings is at least ¼, which is hugely higher than it would be for unrelated individuals. The text entries in the blue parental rectangle in the second bar from the left indicate one of my haplotypes. (b) Population association between HLA types. The close chromosomal juxtaposition of the HLA genes leads to population associations between the HLA types. Thus, the UK population frequency of the variant *A1* of the *HLA-A* gene is 0.178 and the frequency of *B8* at the nearby *HLA-B* is 0.128. If they were independent of each other, the expected frequency of the combination *A1–B8* would be 0.178 × 0.128 = 0.023, whereas the observed frequency is 0.071, more than three times that expected assuming

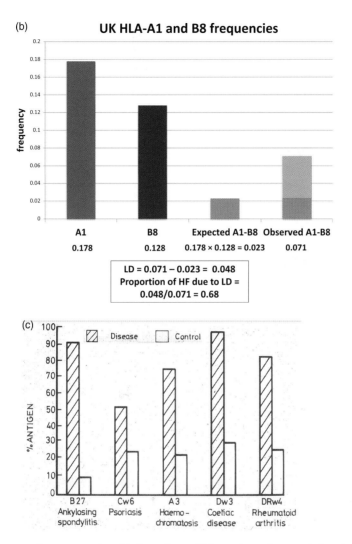

(b) **UK HLA-A1 and B8 frequencies**

	A1	B8	Expected A1-B8	Observed A1-B8
	0.178	0.128	0.178 × 0.128 = 0.023	0.071

LD = 0.071 − 0.023 = 0.048
Proportion of HF due to LD =
0.048/0.071 = 0.68

(c)

FIGURE 5.2 (*cont.*) independence. This tendency for associations between the frequencies of genetic variants at closely juxtaposed genes is called linkage disequilibrium (LD). (c) HLA type associations with five diseases. The bar graphs show the frequencies of HLA types significantly associated with ankylosing spondylitis, psoriasis, haemochromatosis, coeliac disease, and rheumatoid arthritis. In each case the frequency of the particular associated HLA type in diseased patients is compared with that in healthy controls.

language in our genomes and that, on average, any two individuals differ in about three million of these letters, this leaves room for obtaining data on the frequencies of millions of genetic variants for the study of the genetic relationships between human populations.

The HLA model for studying associations between genetic variants and particular diseases, with the aim of discovering inherited genetic contributions to the disease being investigated, can now be extended to any disease or trait using this huge range of DNA sequence variants. All that is needed is to find variants that are found more frequently in people with a disease than in non-diseased controls. For this to work, it is essential to have an appropriate choice of control populations to match the population from which the diseased individuals are obtained for the study. Such studies, called genome-wide association studies (GWASs), are now extensively carried out.

People of the British Isles (PoBI)

The aim of the PoBI project was to make a detailed study of genetic variation throughout the UK and, through that, create a control UK population as a resource for UK-based genetic disease association studies. It was for the latter reason that the UK Wellcome Trust provided more than 10 years of support for the PoBI project, with myself as the leading investigator, joined by my colleague Peter Donnelly, who supervised the analysis of the data (Leslie et al. 2015).

The first requirement was to have a sample of individuals that was truly representative of the different parts of the UK and hence uncompli-cated by recent migrants into those areas. To do this, we aimed to sample volunteers only in rural areas and only those for whom their four grandparents came from approximately the same area, defined as born within 80 km of each other. The aim was then to cover the British Isles by selecting volunteers who represented, as far as possible, the areas from where they were recruited, as defined by these criteria. This criterion of selection has the effect of sampling the DNA of our volunteers' grand-parents. Since the average year of birth of those grandparents was 1885, this amounted effectively to analysing the British population of the nineteenth century before the major population movements associated with agricultural mechanisation and the First World War. Sampling only from rural areas avoids the effects of the extensive, comparatively recent migration following the industrial revolution, largely into towns and cities. As the focus of the PoBI project was on the origins of the major populations of Great Britain and Northern Ireland, the study, intentionally, did not consider the much more recent immigrations to the UK from the Commonwealth and Eastern European countries.

A wide variety of approaches was used to attract volunteers in each of the rural areas we visited. These included substantial help from the local news media, attending agricultural shows, contact with local history and genealogical societies, Women's Institutes, the National Farmers' Union, Rotary International, and branches of the Association of Inner Wheel Clubs and the National Association of Tangent Clubs. In many cases, we also gave talks to attract an interested audience and included a little free refreshment. Probably no more than about 5 per cent of people in the rural areas that were sampled satisfied the criteria set for collection. The basic demographic information obtained from each volunteer included place and year of birth of their grandparents and parents, as well as their own place of residence, gender, and surname at birth. Blood samples were collected as a source of DNA for the genetic studies, and informed consent forms were obtained.

Because of the expectation that any regional genetic differences within the UK would probably be quite small, at least 500,000 DNA sequence variants were analysed. Finally, on the basis of an analysis of the then available HLA data, it was clear that taking into account LD, namely not only the frequencies of the individual genetic variants, but also associations between them based on their relative positions on the chromosome, would give valuable extra information on patterns of genetic variation, rather like the higher resolution that is achieved by using a magnifying glass when studying a map.

The final result, based on analysis of 2,039 volunteers, was a genetic map of the people of the British Isles, as shown in Figure 5.3 (Leslie et al. 2015). The fineSTRUCTURE algorithm (Lawson et al. 2012) used for the analysis of the genetic data creates clusters of individuals based on their overall genetic similarity, so that individuals within a cluster are genetically more similar to each other than they are to the individuals in

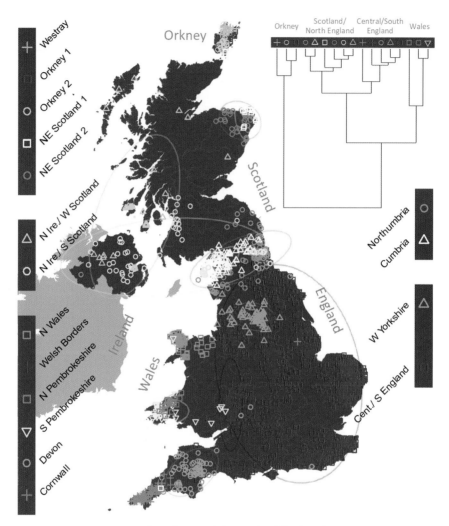

FIGURE 5.3 Genetic map of the people of the British Isles. (Leslie et al. *Nature* 2015 containing Ordance Survey data E Crown copyright and database right 2012, and E EuroGeographics for some administrative boundaries.) Each symbol represents a single individual. Individuals are clustered only on the basis of their genetic similarities, using the fineSTRUCTURE algorithm, and membership of a cluster is indicated by a unique symbol combining colour and shape. Individuals are plotted on a map of the UK at positions corresponding to the mean position of the place where their grandparents were born. No relationship between clusters is implied by the colours and symbols. The broad relationships among the clusters are shown in the top right in the form of an evolutionary tree. As expected, this shows the close relationship between the clusters based in Orkney on the left, next to those in Scotland and northern England, and on the right the Welsh clusters.

any other cluster. The analysis does not just count the number of these positions where two individuals have the same DNA letters, but also takes into account where these positions are along the chromosomes. Thus, whether two positions that are close together on the chromosome both differ, or are both the same, provides very valuable additional information on the extent of similarity between any two individuals at the DNA level. This is the way the algorithm takes LD into account. The fineSTRUCTURE algorithm also allows us to assess the relative extent of differences between people in different clusters. Thus, it can show whether two particular clusters of people are more similar to each other on average than they are to the people belonging to any other cluster.

In the map shown in Figure 5.3, each symbol represents a single individual. Individuals are clustered only on the basis of their genetic similarities, using the fineSTRUCTURE algorithm, and membership of a cluster is indicated by a unique symbol combining colour and shape. Individuals are plotted on a map of the UK at positions corresponding to the mean position of the place where their grandparents were born. No relationship between clusters is implied by the colours and symbols. The most striking result shown by the map in Figure 5.3, which includes 17 clusters of size 10 or greater, is the remarkable correspondence between the genetic clusters and their geographical location. Even Cornwall and Devon, the two neighbouring counties in the southwest corner of England, are quite clearly separated. The broad relationships among the clusters are shown in the top right of Figure 5.3 in the form of an evolutionary tree. As expected, this shows the close relationship between the clusters based in Orkney on the left, next to those in Scotland and northern England, and the Welsh clusters on the right.

The clusters were defined by gradually building up groups of individuals with similar genetic makeups, starting with most similar within and most different from all others, then the next most similar within and different from all the rest, and so on, until further clustering does not improve the explanation of the results. Having done this, we went back 'up the tree' to the first cluster identified and worked down to give a hierarchy of clusters starting from the most different and gradually descending to the least different.

The most different of all the clusters from the rest of the UK is that found in Orkney, which clearly corresponds to the existence of a Norse

Viking earldom in Orkney from 873 to 1468 (Figure 5.4(a)). Jack Rendall, a volunteer who took part in a Channel 4 UK TV series about PoBI, has a highly characteristic Orkney surname. The UK-wide distributions of the Rendall surname in 1881 and 1988 (Figure 5.4(a)) show that the peak area in which the Rendalls live has hardly changed in more than 100 years. On the basis of a simple analysis of Jack Rendall's genetic makeup, he was judged to be more similar to the Welsh (by a factor of 4.4) than to the Norse, but he has a Y (male) chromosome type found most frequently in Orkney, clearly suggesting some male Norse ancestry.

The next most distinctive clusters are the Welsh (Figure 5.4(b)), possibly reflecting the closest relationship to the earliest humans that came to the UK after the end of the last ice age about 12,000 years ago. David Hughes, another volunteer for the TV programme, has a surname with a very obviously Welsh origin. Here again, the UK-wide distributions of the Hughes name in 1881 and 1998 show an amazing stability over 100 years (Figure 5.4(b)). A rough calculation suggests that there is more than 21 times the chance of finding David Hughes' combination of genetic types in Welsh than in Anglo-Saxon (East Anglia/Lincolnshire) populations, a remarkable concordance between the history of the Hughes surname and the genetics.

The next separations distinguish Cornwall from the rest of England and from Northern Ireland, northern England and Scotland, and, most strikingly, north from south Wales. Following this, in stages, come a split in Orkney (perhaps associated with the possibility that the Norse colonists of different parts of Orkney came from different localities in Norway), a separation of Scotland from northern England, and a split of Devon from Cornwall with a remarkably clean separation at the county border. Next, there is a new cluster on the Welsh borders and an intriguing split in South Wales, and finally further splits on the Welsh Borders and in northern England, giving rise to the 17 clusters (Figure 5.3).

It is striking that, on the basis of this hierarchical clustering, north and south Wales are about as distinct genetically from each other as are central and southern England from northern England and Scotland, and the genetic differences between Cornwall and Devon are comparable to or greater than those between the northern English and the Scottish samples.

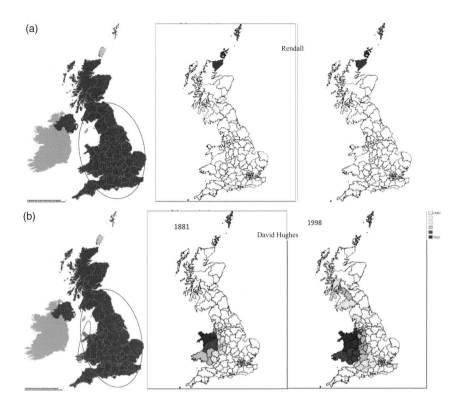

FIGURE 5.4 The genetic geography of surnames. (a) Orkney separates first from the rest of the UK in the Genetic Map and Rendall is an Orkney name. The distinctive separation of Orkney (yellow triangles) from the rest of the UK (red squares) clearly corresponds to the existence of a Norse Viking earldom in Orkney from 873 to 1468. The UK-wide distributions of the Rendall surname in 1881 and 1988 show that the peak area (purple) in which the Rendalls live has hardly changed in more than 100 years (Paul Longley Surname profiler UCL). (b) The next most distinctive clusters in the UK genetic map are the Welsh (pink circles). The UK-wide distributions of the Hughes name in 1881 and 1998 show a remarkable stability over 100 years (purple highest and red next-highest densities, Paul Longley Surname profiler UCL).

The most different of all the clusters from the rest of the UK are those found in Orkney, as already mentioned, which very plausibly results from the Norse Viking Earldom in Orkney from 875 to 1472. The next level of separation shows that Wales forms a distinct genetic group, followed by a further division between north and south Wales. This division corresponds well to the ancient kingdoms of Gwynedd (independent from the

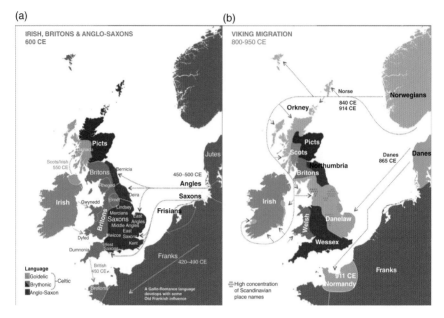

FIGURE 5.5 The peopling of the British Isles in the first millennium AD. (Leslie et al. *Nature* 2015 containing Ordance Survey data E Crown copyright and database right 2012, and E EuroGeographics for some administrative boundaries.) (a) The regions of ancient British, Irish, and Saxon control after the Anglo-Saxon invasions. (Copyright: EuroGeographics for the administrative boundaries (coastlines), Leslie et al. 2015.) (b) The migrations of Norse and Danish Vikings. The main regions of Norse Viking (light brown) and Danish Viking (light blue) settlement are shown. (Copyright: EuroGeographics for the administrative boundaries (coastlines), Leslie et al. 2015.)

end of the Roman period to the thirteenth century) in the north and Dyfed in the south (Figure 5.5(a)).

Subsequently, the north of England, Scotland, and Northern Ireland collectively separated from southern England. Then, at the next level, Cornwall (pink crosses, Figure 5.3) forms a separate cluster quite distinct from Devon (blue squares, Figure 5.3), followed by Scotland and Northern Ireland separating from northern England. The splits in the Northern Ireland group, one with the Scottish Highlands and the other with the lowlands, suggest association with the people of Dalriada and with the Picts, respectively, a separation of clans that existed around 600 AD.

The split in south Wales (pink squares and yellow inverted triangles, Figure 5.3) is suggestive of 'Little England beyond Wales', as will be discussed later. Particularly striking is the distribution of the large cluster of people (red squares) that covers most of eastern, central, and southern England and extends up the east coast. This cluster contains almost half the individuals analysed (1,006). Several of the other genetic clusters show similar locations to the tribal groupings and kingdoms at the time of the Saxon invasion (from the fifth century), suggesting that these tribes and kingdoms may have maintained a regional identity for many centuries. For example, the Cumbrian cluster corresponds well to the kingdom of Rheged, West Yorkshire to the Elmet, and Northumbria to the Bernicia (Figure 5.5(a)). The existence of these quite well separated clusters suggests a remarkable stability of the British people over long periods of time. This is in marked contrast to what is often assumed.

It is important to emphasise that, although the genetic clustering found by fineSTRUCTURE analysis is quite clear and statistically very significant, it is based on very small genetic differences. Nevertheless, it seems very likely that we would be able in many cases to assign an individual of unknown origin to their cluster location. This is illustrated by the fact that an individual was sampled in the northeast of Scotland but was thought to come from Blackburn near Manchester. It turned out, on further checking of the paperwork, that this individual came from a Blackburn that was in Aberdeenshire, and not the better-known Blackburn in Lancashire.

The Impact of the Surrounding European Populations on the Genetic Structure of the British Populations

What is the origin of all this genetic structure in the British populations and where did it come from? To answer this, we need to look to the surrounding European countries, but first, we must ask what can history tell us? The population of the UK has a relatively simple history compared with the rest of Europe. Great Britain was recolonised by modern humans after the last ice age, some 12,000 years ago, with the next major perturbation being the arrival of agriculture about 6,000 years ago. There is uncertainty about the degree of population movement associated with the arrival of Beaker traditions at the end of the Neolithic period about 4,500 years ago, as will be discussed later. The first major land invasion was the Roman occupation from 43 to 410 CE. The evidence indicates that the Romans did not leave many people behind relative to the then size of the British population. Soon after the Romans came the Saxons and related groups, and then the Vikings around 800–950 CE. Finally, there was the Norman invasion of 1066, after which there has been no further major successful land invasion of Great Britain.

In considering the contribution of human migration to changing culture, it is very important not to confuse the influence of relatively small elite groups, such as the Romans and the Normans, who may have a huge impact on culture and language but little impact on the genetics of the masses, with that of larger-scale migrations, such as those of the Anglo-Saxons into England and, in the Scottish Isles, Norse Vikings. History and, to a lesser extent, archaeology generally focus on the elite groups, whereas genetics looks at the constitution of the whole of the population, including the peasants on the land who, after agriculture, were the main mass of the population until comparatively recently. Of all the major migrations to Great Britain after its initial settlement some 12,000 years ago, the historical and archaeological evidence suggests that it is only, possibly, the Neolithic, and clearly the Anglo-Saxon (Figure 5.5(a)) and Viking invasions (Figure 5.5(b)) that are likely to have had a major impact on the genetics of the indigenous population of the British Isles. To pursue the genetic effects of these major invasions, we have to study the genetic relationships between the UK populations and those of the surrounding European countries. For this, a study of European populations along the lines of the PoBI project is needed in order to assess the extent to which the European countries have contributed to the genetic composition of the British genetic clusters.

This need was, fortunately, satisfied by a contemporary study of the genetics of multiple sclerosis based on 6,209 individuals from 10 different European countries (Sawcer et al. 2011). The genetic data obtained for this study, using essentially the same set of genetic markers as for the PoBI study, were first analysed into 51 genetic clusters (excluding Ireland) using the fineSTRUCTURE algorithm just as for the British samples (Figure 5.6(a)). In Figure 5.6(a), each cluster is given a unique number and colour: the sizes of the circles are in proportion to the number of samples collected from where they are located; and the relative sizes of the coloured sectors within circles are proportional to the numbers of individuals found at that location that belonged to the corresponding cluster. The sourcing of the European samples was much more coarse-grained than that for PoBI as the multiple sclerosis study did not limit samples to rural areas or impose the four-grandparent condition. Thus, some of the data from large cities, such as Berlin or Stockholm, behaved as if they were countries composed of a mixture of several clusters.

Most European countries were well separated into different genetic groups by this analysis, with some showing significant internal genetic heterogeneity, analogous to that found in the UK. Norway, Sweden, Finland, and Denmark were remarkably, and unexpectedly, well separated from each other. Italy was well separated from the northern European countries, and Poland was well separated to the east of Germany. Germany was split into northwest, northeast, and southern clusters, while France had a southern cluster that overlapped with Spain.

The next question was how could one relate these European clusters to the UK clusters? The answer was to use a complex statistical regression analysis (Leslie et al. 2015) that estimated the representation of each of the British clusters as a mixture of different proportions of the European clusters. Counting Norway as a single source, only 9 of the 51 European clusters identified by the fineSTRUCTURE analysis contributed significantly to the British clusters. These were clusters 12, 14, 17, and 31 in France, 11 in Belgium, 18 in Denmark, 3 and 6 in Germany, and finally Norway, all representing populations at the seaside of the countries surrounding Great Britain, just as might be expected. A summary of the results of this analysis is shown in Figure 5.6(b).

In Figure 5.6(b), the most obvious European cluster contribution that fits expectations is the bright blue of Norway. Its largest contributions are to the three Orkney clusters, with somewhat lower contributions to northern Scotland and Northern Ireland, and gradually decreasing contributions moving southwards, to a very low level in the large central, eastern, and southern England group (red squares in Figure 5.3). This matches what is expected from the Norse Viking settlements. The average Norwegian contribution to the Orkney samples represents only about a 25 per cent Norse Viking admixture. This clearly shows that

(a)

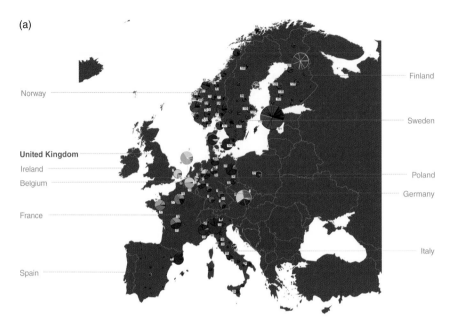

FIGURE 5.6 The distribution of genetic clusters in mainland Europe and their contribution to the British Isles. (a) Genetic map of mainland European countries. There are 51 clusters. Each cluster is given a unique number and colour, the sizes of the circles are in proportion to the number of samples collected from where they are located, and the relative sizes of the coloured sectors within circles are proportional to the numbers of individuals found at that location that belonged to the corresponding cluster. (Leslie et al. 2015, containing Ordance Survey data E Crown copyright and database right 2012, and E EuroGeographics for some administrative boundaries.) (b) British Isles clusters represented as mixtures of different proportions of European clusters. Each pie chart represents one of the 17 British clusters indicated in Figure 5.3, and the relative contributions of the different European groups to that cluster are proportional to the sizes of the sectors in the pie charts, with the colour of the sector indicating its source. Only 9 of the 51 European clusters contributed significantly to the British clusters. These were clusters 12, 14, 17, and 31 in France, 11 in Belgium, 18 in Denmark, 3 and 6 in Germany, and Norway, colour coded as indicated in the figure.

the Norse Vikings certainly did not wipe out the resident Pictish population and replace it, but rather intermarried significantly with it.

The three Welsh clusters are the most distinctive and completely lack contributions from north and northwest Germany (EU3 pink) and northern France (EU17 red). They have the largest contributions from west

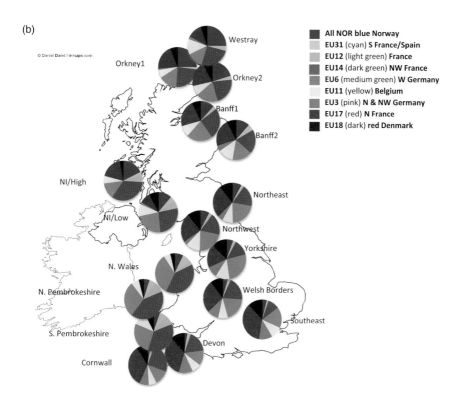

(b)

© Daniel Dalet / d-maps.com

Westray
Orkney1
Orkney2
Banff1
Banff2
NI/High
Northeast
NI/Low
Northwest
Yorkshire
N. Wales
N. Pembrokeshire
Welsh Borders
S. Pembrokeshire
Southeast
Devon
Cornwall

All NOR blue Norway
EU31 (cyan) **S France/Spain**
EU12 (light green) **France**
EU14 (dark green) **NW France**
EU6 (medium green) **W Germany**
EU11 (yellow) **Belgium**
EU3 (pink) **N & NW Germany**
EU17 (red) **N France**
EU18 (dark) **red Denmark**

FIGURE 5.6 (*cont.*)

Germany (EU6 medium green) and northwest France (EU14 dark green). This configuration strongly suggests that the Welsh may be closest to the original settlers who came to Britain after the end of the ice age. While there is no clear 'Celtic Fringe', as is so often assumed, there is evidence of ancient British DNA in common with other British populations, especially in Scotland and Northern Ireland, but less in Cornwall or Devon, in contrast to what might have been expected. The France cluster 12 (light green) is intriguingly found only in Wales and Northern Ireland and could possibly represent an early migration that came from farmers from the western route, along the Atlantic coast, rather than across the English Channel.

The small differences between south and north Pembrokeshire, especially the slightly larger contributions from Belgium (EU11 yellow) and

Denmark (EU18 dark red), matching Danish place names in south Pembrokeshire, are consistent with the suggestion that this group may represent the area that is sometimes called 'Little England beyond Wales'. The suggestion is that this difference originates from Flemish farmers settled there by Henry I in the early twelfth century. Their descendants kept to themselves and kept out the Welsh speakers creating the Landsker line, a language border in Wales between the largely Welsh-speaking area and what became the largely English-speaking area.

The most obvious contribution representing the Anglo-Saxons is EU3 (pink) from north and northwest Germany. That is consistent with the lack of evidence for Anglo-Saxon incursions into Wales. Denmark (EU18, dark red) is another candidate for an Anglo-Saxon contribution. On the basis of these two contributions, the best estimates for the proportion of presumed Anglo-Saxon ancestry in the large eastern, central, and southern England cluster (red squares in Figure 5.3) are a maximum of 40 per cent and could be as little as 10 per cent. This is strong evidence against an Anglo-Saxon wipe-out of the resident ancient British population, and clearly indicates extensive admixture between the incoming settlers and the indigenous people. The difference between Devon and Cornwall is most probably due to the greater Saxon influence in Devon, this being consistent with the slightly greater contributions of EU3 (pink) and EU18 (dark red) to the makeup of the Devon cluster compared with the Cornwall cluster.

The homogeneity of the eastern, central, and southern British cluster (red squares in Figure 5.3), with no obvious differences in the Danish contribution (EU18, dark red) between it and the more northern English populations (where any Viking contributions in the northwest of the region covered by the cluster would have come from Norwegians), suggests that the Danish Vikings, in spite of their major influence through the 'Danelaw' and many place names of Danish origin, contributed relatively little of their DNA to the English population. There may also be a residual contribution to this homogeneity from disruption to the Iron Age tribal entity by the Roman occupation, which covered more or less the same area as this large cluster.

The most intriguing, and novel, European contribution to the British clusters is that from northern France (EU17, red). This clearly post-dates the original settlers, since it is entirely absent from the Welsh samples. It is, however, widespread elsewhere, even through the north of England and Scotland to Orkney. It is also especially prevalent in Cornwall and Devon. This distribution suggests a substantial migration across the channel after the original post-ice-age settlers but before Roman times. Barry Cunliffe points out that even by 1500 BC 'this ever-increasing mobility had created broad cultural similarities along both sides of the eastern Channel–south North Sea interface' and 'In the second half of the second millennium BC the communities of south-eastern Britain were in very close contact with those living on the adjacent coasts of northern France, to such an extent that much of the material culture and settlement archaeology on both sides of the Channel developed close similarities.' (Cunliffe 2013). This sounds just like a description of the northern France (EU17, red) contribution to the UK clusters.

Overall, the European clusters that make relatively large contributions to all the UK clusters are west Germany (6, medium green), northwest France (14, green), and Belgium (11, yellow), which suggests that these are the best representatives of the earliest settlers to Great Britain, consistent with the Welsh being the closest current population to the early settlers. There is also evidence for only a very small Spanish contribution to all the PoBI clusters, in contrast to what has been claimed by some authors.

Ireland

PoBI was not funded to study Ireland as the Wellcome Trust mainly supports such research in the UK. More recent studies have now been carried out in Ireland, which allow a comparison between the UK and Ireland by doing an analysis on the combined data. Using an overlapping set of genetic variants and the same fineSTRUCTURE approach to analysis gave rise to a genetic fine structure map of Ireland comparable to that for Great Britain (Figure 5.7(a), Gilbert et al. 2017). This shows genetic clusters that mostly fit into the four Irish provinces, with only

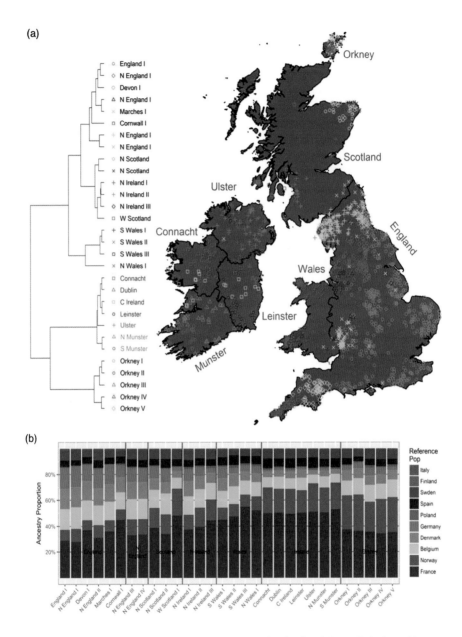

FIGURE 5.7 Fine genetic structure of Ireland and Great Britain and its composition in terms of mixtures from mainland Europe (adapted from Gilbert et al. 2017, Figure 1). (a) Genetic fine-structure map of Ireland compared with that of Great Britain. As in Figure 5.3, each symbol represents a single individual, individuals are clustered only on

two of the three major clusters (blue crosses and red diamonds) in Ulster, corresponding to Northern Ireland, overlapping with the UK in Scotland and to a limited extent in northern England, effectively matching and extending what was found in the PoBI analysis. The third major genetic cluster (purple crosses) in Ulster was concentrated to the south of the province and was the only group that overlapped with the Republic of Ireland. The phylogenetic tree (on the right of Figure 5.7(a)) connecting the Irish and UK genetic clusters shows that the strongest affinity between the Irish and the UK samples (including the purple crosses in the map of Ulster with Ireland) is with Orkney.

The relationship of the Irish clusters to European populations was investigated by an analysis of the contributions of European genetic clusters to the Irish clusters following the same approach as described above for PoBI. The results of this analysis are shown in terms of broad contributions by European countries to the various UK and Irish genetic clusters (Figure 5.7(b)). For the non-Irish samples, the results are in general agreement with those described above for the UK, with relatively large contributions from Germany (dark green), Denmark (blue), and Belgium (light green) to southern England, and the largest contribution from Norway (purple) to Orkney. Most striking, however, is the uniform and relatively high Norwegian contribution to all the Irish samples, apart from those that overlap the UK. This suggests a much larger contribution to overall Irish ancestry from the Norse Viking invasions than was suspected from earlier, less comprehensive analyses. This pervasive Norse Viking contribution to Irish ancestry probably accounts for the relationship between the Irish populations and the Orcadians shown by the phylogenetic analysis (Figure 5.7(a)).

CAPTION FOR FIGURE 5.7 (*cont.*) the basis of their genetic similarities, using the fineSTRUCTURE algorithm, and membership of a cluster is indicated by a unique symbol combining colour and shape. Individuals are plotted on a map at positions corresponding to the mean position of the place where their great-grandparents or grandparents were born. The phylogenetic tree (on the left) connects the Irish and UK genetic clusters (Gilbert et al. 2017, Figure 1). (b) Contributions of major European countries to each of the 30 Irish and British clusters. The contributions of each of the 10 European countries are given by the size of the vertical segment for each of the 30 clusters according to the colour code on the bottom right.

Neolithic Migrations and Ancient DNA Studies

The development of the agriculture that spread to Europe started in the fertile crescent in the Middle East around 10,000 years ago. It then spread gradually throughout Europe over the next 5,000 to 6,000 years, reaching Great Britain just over 6,000 years ago. Once having reached the British Isles, agriculture spread relatively rapidly throughout Great Britain and Ireland. A key question is whether the spread of agriculture was cultural, essentially by word of mouth, or by movements of the farmers, referred to as 'demic' spread. Cavalli-Sforza and colleagues seeking an answer to this question showed that there were consistent coordinated gradients of genetic variant frequencies across Europe, spreading from somewhere in the Middle East out to the northern periphery of Europe (Menozzi et al. 1978). They argued that this was consistent with a gradual advance of farmers moving into new territory and mixing with their new neighbours. This would lead to a gradual dilution of any differences in genetic variant frequencies between the original farmers of the Middle East and the populations into which they were advancing, giving rise to just the sort of frequency gradients Cavalli-Sforza and colleagues had observed. The essential factors determining the gradient are (1) the existence of genetic variation in the area covered by the expansion of farmers, prior to expansion; (2) the capacity of farmers to grow to a higher population density than the hunter/gatherers; (3) a relatively slow migration rate; and (4) the occurrence of acculturation of hunter/gatherers so that they became farmers. By the time farming reached Great Britain the differences in genetic variant frequencies between the arriving farmers and the original farmers from the fertile crescent would have been diluted out, while the speed with which farming spread throughout Great Britain suggests that this was possibly largely by cultural rather than demic spreading.

There are two main possible routes for the arrival of agriculture to Great Britain. The first, and seemingly most direct, is from Anatolia directly across the centre of Europe. The second is from the Aegean following the Atlantic coast of the Iberian Peninsula and then by boat across the channel. As Cunliffe says, 'There are, then, two clear narratives that can be constructed to explain the introduction of the Neolithic way of life to Britain and Ireland, an east-side story of intensive links between north-eastern France and the eastern coasts of Britain, and a west-side story of more dispersed movements along the Atlantic seaways.' (Cunliffe 2013, 144–145).

Archaeologists have described the Beaker phenomenon as a major later culture that spread across Europe arriving in Great Britain from northern France in about 2500–2200 BC. The characteristic Beaker pottery is commonly found with burial sites. There is also a suggestion that the Beaker cultural change may have overlapped to some extent with a movement of semi-nomadic peoples, called the Yamnaya, originating mainly from the European steppes, who also had a distinctive burial culture.

The recent development of techniques for obtaining DNA sequences from DNA extracted from skeletons from burial mounds dating back nearly 6,000 years, to when agriculture first arrived in Great Britain, has made it possible to analyse the genetic variants found in these ancient people. This has naturally led to the question of how the genetic variants found in these ancient individuals compare with those studied in contemporary populations, such as in the PoBI project. The experimental procedures involved for the analysis of ancient DNA are, as might be expected, technically very demanding and so give results that are mostly not yet of the same quality as for modern DNA, creating some difficulties for the comparison of ancient and modern genomes.

Two recent papers analysing the DNA of ancient British genomes have made striking claims that the comparisons of these genomes with their modern counterparts show evidence for a total replacement of the indigenous peoples of Great Britain around 4000 BC by incoming farmers from the Aegean (Brace et al. 2019) and then another total replacement of these people by the later incomers, starting around 2500 BC, who brought the Beaker culture (Olalde et al. 2018). They drew these conclusions on the assumption that the skeletons from which they obtained their DNA for genetic analysis, which came from graves throughout Great Britain that matched, respectively, the times of the initial influx of the farmers, or of the Beaker people, were true representatives of the then major inhabiting populations of the areas where the graves were found.

Thus, Brace et al. (2019) state that 'Our analyses indicate that the appearance of Neolithic practices and domesticates in Britain circa

4000 BC was mediated overwhelmingly by immigration of farmers from continental Europe and strongly reject the hypothesised adoption of farming by indigenous hunter-gatherers as the main process' and, further, that 'Genetic affinities with Iberian Neolithic individuals indicate that British Neolithic people were mostly descended from Aegean farmers who followed the Mediterranean route of dispersal.'

In an earlier paper, from an overlapping group of workers, Olalde et al. (2018) state that 'The arrival of people associated with the Beaker Complex precipitated a profound demographic transformation in Britain, exemplified by the presence of individuals with large amounts of Steppe-related ancestry after 2450 BCE' and that this 'was associated with the replacement of approximately 90% of Britain's gene pool within a few hundred years'.

There are two major problems with these conclusions. The first is that such major population replacements by immigrants, especially over relatively short periods of time, are a priori unlikely. The second is that, unless even a third population replacement is assumed, the incoming 'Beaker' population should show a reasonable relationship to the current population of the UK as analysed by the PoBI project.

Since the ratio of immigrants to an indigenous population is generally quite low, major population replacements by immigrants depend either on a high rate of increase of the immigrants relative to the indigenous population, or on slaughtering the indigenous population, or a mixture of these. The expansions of Europeans into America and Australia are obvious examples of major replacement involving significant killing of the indigenous population. Nevertheless, in Mexico and in South America, for example, there remains a strong genetic component of native American ancestry in spite of the killing and differential disease susceptibility. Even the marauding Norse Vikings and Anglo-Saxons only ended up contributing not much more than 25–30 per cent to the eventual genetic makeup of the Orkney and southeast English populations.

The suggestion of two successive replacements creates even more of a problem. The Beaker people are assumed to have replaced the earlier Neolithic peoples within about 1,500 years of each other. The time the Beaker people came to Great Britain was therefore more than 1,000 years after farming had arrived. By that time, there would probably already have been a substantial increase in the then indigenous British population due to the uptake of farming, making complete replacement of the indigenous, already Neolithic population by the incoming Beaker people very hard to achieve. If the Beaker people really replaced up to 90 per cent of the then indigenous population, and there was no further such complete displacement, which surely must be assumed to be the case, then the Beaker people's genetic signature should overlap significantly with that of at least some parts of the current UK.

The PoBI analysis, which takes into account the much later Anglo-Saxon and Viking invasions, clearly indicates that the current Welsh population is likely to be the closest of the current UK populations to what would have become the indigenous British population after arrival of the Beaker people. The results of Olalde and colleagues' analysis did not, however, suggest that the Beaker skeletons they had analysed from Great Britain had any obvious relationship to current UK populations, though they did not make direct comparisons with the PoBI population data. They suggested, rather, that the British Beaker people were related to those from the European steppes from which some of the original Beaker populations were assumed to have come.

The analysis of early British farmers by Brace et al. (2019) suggested they were descended from Iberian Neolithic-related populations, Cunliffe's 'west-side story' discussed above, with no relationship to the current Welsh population. This would imply, if these Iberian Neolithic farmers had made a substantial contribution to the indigenous British population, a substantial Spanish contribution to the current populations, for which there is no evidence. This lack of a Spanish contribution would, of course, be expected, if the Beaker people had more or less completely replaced their predecessor Neolithic population.

There is one further interesting slightly earlier study by Galinsky et al. (2016), which compares ancient DNA from Anglo-Saxon samples from Britain, Neolithic Anatolian samples, and certain other ancient samples from mainland Europe directly with modern British samples. This study uses extensive data on 113,851 samples from the 500,000 samples of the Great Britain-wide BiobankUK study.[1] The samples used exclude

---

[1] www.ukbiobank.ac.uk.

relatively recent immigrants from British Commonwealth countries and Eastern Europe in order to focus on the original British population. The results of these British samples were matched with those from PoBI and the ancient DNA samples. The BiobankUK study is much larger than PoBI (which was originally proposed as a control for BiobankUK) but has less precise information on the geographical origin of the participants and no focus on ancestry. An extract of the results of Galinsky and colleagues' study, based on principal-components analysis (PCA), is illustrated in Figure 5.8. PCA is a widely used method, but less precise than fineSTRUCTURE, for clustering populations of individuals according to their genetic similarity. In Figure 5.8, each dot represents an individual, and the colours indicate where an individual comes from. Thus, in the two squares on the left, purple is northern England, blue is southern England, brown is Northern Ireland, green is Scotland, red is north Wales, and orange is south Wales. The top square (PC1 against PC2) separates all the populations except for north and south Wales. The bottom square (PC2 against PC5), on the other hand, gives a clear separation between north and south Wales (red and orange), but with Northern Ireland (brown) and Scotland (green) overlapping. In the two central squares, the coloured spots represent PoBI samples using the same colour coding for their origin as above. The PoBI samples fall exactly where expected, though the overlap between the brown (Northern Ireland) and green (Scotland) in the lower square is harder to see because there are relatively fewer samples from Northern Ireland. This shows that, within the limits of this approach to analysis, there is very good agreement between the BiobankUK and PoBI samples although they were collected in quite different ways. The fine structure detected in PoBI is necessarily not revealed by the BiobankUK analysis.

The two squares on the right in Figure 5.8 indicate where the ancient DNA samples land on the genetic map of the modern samples. Here, Anatolian Neolithic farmers are indicated in yellow, Mesolithic Europeans in blue, Yamnaya steppe dwellers in brown, and Saxons from east England in blue. Only the Anatolian Neolithic farmers stand out as different from the modern British samples. Their difference, according to Galinsky et al. (2016), matches what might be expected from a significant

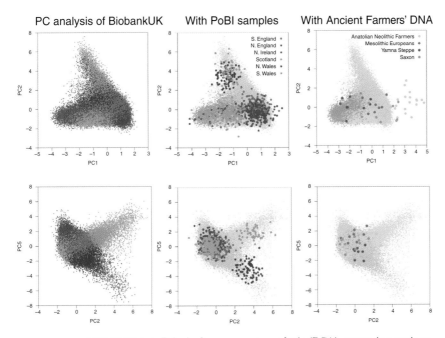

FIGURE 5.8 Principal-components analysis (PCA) comparing ancient DNA and People of the British Isles (PoBI) samples with those from the UK wide BiobankUK study. Each dot represents an individual, and the colours indicate where an individual comes from. In the two squares on the left: purple is northern England, blue is southern England, brown is Northern Ireland, green is Scotland, red is north Wales, and orange is south Wales. In the two central squares, the coloured spots represent PoBI samples using the same colour coding. The two squares on the right show where the ancient DNA samples land on the genetic map of the modern samples. Here, Anatolian Neolithic farmers are indicated in yellow, Mesolithic Europeans in blue, Yamnaya steppe dwellers in brown, and Saxons from east England in blue (Galinsky et al. 2016, adapted from Figures 1, 2, and 3).

Spanish contribution, just as expected for the farmers from the west. That this is outside the map for the UK samples is expected because of the small Spanish contribution detected in the UK samples by the PoBI analysis. All the other ancient DNA signals are consistent with the much later admixture that came with Vikings, Anglo-Saxons, and the 'Cross-Channel' EU17 red group. In particular, the placing of the English Saxon samples fairly and squarely within the map of the modern UK samples shows that, at least at that time some 1,500 years ago, there was not a huge difference between ancient and modern populations. However, it is notable that none of the ancient DNA samples maps onto the Welsh (see Figure 5.8, the bottom right-hand square).

Conclusion

So, given that the ancient DNA results seem to be in conflict with the analysis of current populations, and that two major population replacements, first by the farmers and then by the Beaker people, seem most unlikely, what can be the explanation? I suggest that the most likely possibility is that the burial sources of the ancient DNA samples do not reflect the overall population that existed in Great Britain at that time. This possibility was considered by Olalde et al. (2018) but rejected because 'their samples are dispersed geographically, extending from England's Southeastern peninsula to the Western Isles of Scotland, and come from a wide variety of funerary contexts (rivers, caves, pits, barrows, cists and flat graves) and diverse funerary traditions (single and multiple burials in variable states of anatomical articulation)'.

There are, however, many examples of elite immigrant groups having major social and cultural influences but leaving behind a minimal genetic trace, notably the Romans and the Normans in the UK. The Norman cultural influence was huge while their genetic input probably amounted to no more than 2 per cent of the total population (Cunliffe 2013, 487). A similar argument applies to the uptake of languages, for example the transition from Celtic to Anglo-Saxon/English following at most a 30–40 per cent 'replacement' of the indigenous population's genetic makeup by that of the Anglo-Saxon incomers. Cunliffe points out, in relation to the Beaker people, that 'In Britain the appearance of Beaker burials, with their characteristic array of grave goods, marks a sudden change in culture' and that 'The way in which communities dispose of their dead reflects their deeply held beliefs and their attitudes to the gods.' (Cunliffe 2013, 208, 266). He also points out that cemeteries, which served local communities for many generations, could only cope with a small percentage of the local population, the rest being disposed of elsewhere (Cunliffe 2013, 266). It was generally only the people of high status who were buried ceremoniously.

I conclude, therefore, that by far the most likely explanation for the results of the ancient DNA studies published so far is the one rejected by Olalde et al. (2018), namely that the burial sites from which the ancient DNA samples are obtained are distinctive and were largely for elite subgroups of the incoming populations of farmers or Beaker people that mostly kept to themselves and therefore do not represent the bulk of the indigenous population at that time. The Anatolian Neolithic farmers who came from the west, along the Iberian and French Atlantic coasts, over a period of at least 2,000 to 3,000 years will surely have acquired significant Spanish and French genetic input. Assuming limited intermarriage between the incoming farmers and the indigenous hunter gatherers, this could account for the low level of the Spanish/French cluster (EU 31) in the Welsh and the distinctive French cluster (EU 12) found only in the Welsh. Input from the eastern farmers crossing the channel could similarly account for the low level of Belgian (EU 11) and Danish (EU18) input to the Welsh (Figure 5.6(b)). In general, however, it is still likely that the spread of agriculture in Great Britain was largely cultural and not demic. While a significant input from migrating Neolithic farmers cannot be excluded, this fits the overall low Spanish contribution to the UK population only if there was little admixture between the migrating Neolithic farmers and the endogenous hunter-gatherers.

Only more samples, carefully provenanced, no doubt accompanied by improved DNA technology, will eventually answer these questions conclusively.

Acknowledgements

My thanks go to all our volunteers, without whom we could not have done the PoBI study, and to the Wellcome Trust for their support. Thanks also to all the members of the PoBI group past and present, especially Bruce Winney and Tammy Day, who did so much for organising the sample collections. Also major thanks are due to the analysis team led by Peter Donnelly, especially Stephen Leslie and Garrett Hellenthal, who did most of the analysis, and to Mark Robinson and Barry Cunliffe for their incisive archaeological advice.

References

Brace, S., Diekmann, Y., Booth, T. J., van Dorp, L., Faltyskova, Z. et al. (2019) 'Ancient genomes indicate population replacement in Early Neolithic Britain'. *Nature Ecology and Evolution* 3, 765–771.

Cavalli-Sforza, L. L., and Edwards, A. W. F. (1965) 'Analysis of human evolution'. In S. J. Geerts (ed.) *Genetics Today: Proceedings of the 11th International Congress of Genetics.* Oxford: Pergamon Press, vol. 3, pp. 923–933.

Cunliffe, B. W. (2013) *Britain Begins.* Oxford: Oxford University Press.

Galinsky, K. J., Loh, P.-R., Mallick, S., Patterson, N. J., and Price, A. L. (2016) 'Population structure of UK Biobank and ancient Eurasians reveals adaptation at genes influencing blood pressure'. *American Journal of Human Genetics* 99(5), 1130–1139.

Gilbert, E., O'Reilly, S., Merrigan, M., McGettigan, D., Molloy, A. M. et al. (2017) 'The Irish DNA Atlas: Revealing fine-scale population structure and history within Ireland'. *Scientific Reports* 7, article 17199.

Hirschfeld, L., and Hirschfeld, H. (1919) 'Serological differences between the blood of different races: The result of researches on the Macedonian front'. *Lancet* 2, 675–679.

Landsteiner, K. (1900) 'Zur Kenntnis der antifermentativen, lytischen und agglutinierenden Wirkungen des Blutserums und der Lymphe'. *Centralblatt für Bakteriologie, Parasitenkunde und Infektionskrankheiten* 27, 357–362.

Lawson, D. J., Hellenthal, G., Myers, S., and Falush, D. (2012) 'Inference of population structure using dense haplotype data'. *PLoS Genetics* 8, article e1004224.

Leslie, S., Winney, B., Hellenthal, G., Davison, D., Boumertit, A. et al. (2015) 'The fine-scale genetic structure of the British population'. *Nature* 519, 309–314.

Menozzi, P., Piazza, A., and Cavalli-Sforza, L. (1978) 'Synthetic maps of human gene frequencies in Europeans'. *Science* 201(4358), 786–792.

Olalde, I., Brace, S., Allentoft, M. E., Armit, I., Kristiansen, K. et al. (2018) 'The Beaker phenomenon and the genomic transformation of northwest Europe'. *Nature* 555, 190–196.

Sawcer, S., Hellenthal, G., Pirinen, M., Spencer, C. C. A., Patsopoulos, N. A. et al. (2011) 'Genetic risk and a primary role for cell-mediated immune mechanisms in multiple sclerosis'. *Nature* 476, 214–219.

van Dijk, J., and Longley, P. A. (2020) 'Interactive display of surnames distributions in historic and contemporary Great Britain'. *Journal of Maps*, 16(1), 68–76.

von Dungerne, E., and Hirschfeld, L. (1910) 'Uber Vererbung gruppenspezifischer Strukturen des Blutes'. *Zeitschrift für Immunitätsforschung und experimentelle Therapie* 6, 284–292.

6 Heroes and Villains of Blood

ROSE GEORGE

It is fitting that the story of blood, that abundantly rich substance and subject, should yield endless heroes and villains. Blood itself, for a start, is a substance with an innate dual nature. It can kill us or save us; it can be a lethal weapon (so much so that in some parts of the world you can be jailed for not revealing your HIV status to a sexual partner, even when the HIV is untransmissible) or a remedy so powerful, patients sometimes don't think they are being properly treated without a bag of packed red blood cells hanging at their bedside. In the 1950s, the director Frank Capra made a series of educational films for Bell Laboratories. They included films on the sun, the weather, heredity, the oceans, and time. The second of the series is my favourite, because it is *Hemo the Magnificent*, a partly animated hour-long introduction to the wonders of the human circulatory system. Its hero is a cartoon he-man named Hemo, who is blood; the cast includes cartoon animals that live with Hemo in a forest and two human actors named Mr Fiction Writer and Dr Research. The science is lucidly conveyed, the concept is satisfyingly batty, but I also love it for the poetry, for this is what Hemo says in scorn and anger to the humans:

> Humans think blood means disease, wounds, pain. These friends [the animals], they know me for what I really am: health, life. I'm the song of the lark, the blush on the cheek, the spring of the lamb. I am the precious sacrifice ancient man offered up to his gods, I am the sacred wine in the silver chalice. Down through the ages I am the price men pay for freedom. But to you scientists, I am a smear on a slide, a stain, a specimen, a sickness. My story is a song only poets should sing, not disease-lovers.[1]

[1] From Frank Capra, *Hemo the Magnificent* (1957).

The poets did sing the song of blood, even the ancient ones. What could be more alluring than this substance that, when it appeared, mostly signalled death or grievous injury, yet women could emit the same liquid (apparently) and not only not die, but give life? Perhaps the best example of this enduring puzzle was Medusa, a beautiful woman who was raped by Poseidon, then punished for being raped by Athena, who made her monstrous, with a head of snakes and a petrifying gaze. Medusa is the best known of the Gorgon sisters, and her most commonly told story is of her death brought about by Perseus. But the ancient poets who wrote of her also celebrated her blood, because when Asclepius, the god of medicine, was given blood from her dead body, he used the blood from her left side to take life, and the blood from the right side to raise humans from the dead. She could be our first villain, Medusa, yet she is not. Instead, she stands for what blood is: a mystery, still, with all our knowledge, that often defies categories. So, I will introduce my first real villain in this series with the understanding that he is no villain unless you were his rival, or a seventeenth-century dog, lamb, or cow.

Villain 1: Jean-Baptiste Denis

Jean-Baptiste Denis was a brilliant medical man. He qualified first as a Bachelor of Theology, then studied medicine in Montpellier. By 1667, he was a Professor of Philosophy and Mathematics in Paris, an eminent scientist, and a personal physician of King Louis XIV. That year, he wrote a letter that was published in *Philosophical Transactions,* a journal of the Royal Society, which was a carefully written thumbing of his nose at his peers in London. By now, a set of scientists on either side of the Channel were competing to transfuse blood into a human. It wasn't human blood. Animal blood looked much like human blood and was possibly superior, as blood was thought to transmit character as well as life, and blood from a 'mild and laudable' animal such as a lamb or a 'placid beast', such as a cow, could convey serenity and quiet along with red blood cells. In London, the leading transfusionist was Richard Lower; his counterpart was Denis, and both had spent years doing terrible things to animals in the name of science. 'We have transfused the blood of three Calves into three Dogs', wrote Denis, 'to assure our selves, what the mixture of two such differing

FIGURE 6.1 Transfusion dog. From Johann Sigismund Elsholtz's *Clysmatica Nova* (1665). Wellcome Collection (public domain mark).

sorts of blood might produce.' It should have produced haemolytic diffi-culty or haemolytic shock, which we now know happens when incompat-ible bloods meet. But the dogs were lucky: all survived, 'and one of the three Dogs, from whom the day before so much blood had been drawn, that he could hardly stir any more, having been supplied the next morning with the blood of a Calf, recovr'd instantly his strength, and shew'd a surprising vigor' (Denis 1666a; Lower 1665–1666). Illustrations of these experiments show spread-eagled dogs (Figure 6.1) and a lamb that looks rightly furious rather than mild or laudable.

Denis's letter signalled only the latest round in a battle of the blood transfusers: the year before, Thomas Coxe (1666) had published a paper on 'Bleeding a Mangy into a Sound Dog', and found that the Sound Dog was unaffected, and the Mangy Dog 'in about ten days . . . perfectly cured'.

Sound Dogs were good, but the goal of these transfusionists was a successful human transfusion. First Denis infused a young lad who was wasting away, and who survived and thrived despite his body having to deal with lamb's blood on top of whatever was ailing him. The next subject was an older man, who also survived. Denis's most famous case, though, was his last. Antoine Mauroy, a madman, wife beater, and former valet to nobility, was given the blood of a calf to calm the man's 'phrensy' (Denis 1666b). Despite science knowing nothing about blood types or incompatibility, or much about the nature of blood, Mauroy's body at first did not react adversely. The second transfusion was larger, and Denis described, without knowing, haemolytic shock: 'As soon as the blood began to enter into his veins, he felt the like heat along his Arm ... his pulse rose presently, and soon after we observed a plentiful sweat all over his face.' He vomited up bacon and fat, but the next morning woke calm. 'He made a great glass of Urine, of a colour as black, as if it had been mixed with the blood of chimneys' (it was not soot but his dead cells killed by the foreign blood). Denis claimed success and triumph, but Mauroy soon died. In London, the transfusionists raced to catch up, also picking a madman for their most famous case. Arthur Coga was the brother of the master of Pembroke College, but, in the words of Samuel Pepys (1667), 'crack'd in the head', and a drinker. Richard Lower and his colleagues gave Coga 20 shillings in return for transfusing him with 12 ounces of sheep's blood. Coga survived, but the triumph was short: too many were sceptical of the wisdom of xenotransfusion (the interspecies transfusion of blood), and, as the transfusionist Geoffrey Langdon Keynes (1922) later wrote, there were fears that 'terrible results, such as the growth of horns, would follow the transfusion of an animal's blood into a human being'. In Paris, Mauroy's wife was probably executed for his murder (the record is unclear), and Denis was disgraced. The French outlawed transfusion entirely, and in Britain it fell from favour until the nineteenth century, when a courageous obstetrician named James Blundell wanted to stop his patients dying in childbirth. He could have used animal blood, but had his own delightful reasons for preferring humans to donate.

> What then was to be done on an emergency? A dog, it is true, might have come when you whistled, but the animal is small; a calf, or sheep, might, to some, have appeared fitter for the purpose, but then it could not run up stairs.
>
> (Blundell 1834)

Blundell was partly successful, enough for the way to be paved for transfusion to flourish in the next century, once the Austrian chemist Karl Landsteiner had discovered blood groups. And the Sound Dogs and laudable lambs were left alone, at least by the transfusionists.

Hero 1: Dame Janet Maria Vaughan

To write about Janet, I must first write about Percy. Mr Percy Lane Oliver, born in Cornwall but a transplant to London, worked as a civil servant for Camberwell council. He lived in Peckham with his wife, Ethel Grace, and believed strongly in the need for volunteerism, enough that he had been awarded the Order of the British Empire (OBE) for his services to refugees during the First World War. In 1921, Percy Oliver was 43 years old and honorary secretary to the Camberwell Division of the British Red Cross. Today in 2021 the UK carries out 2.5 million blood transfusions a year, and *nearly a million people regularly donate their blood.*[2] One hundred years ago, there was little sign of the smooth, huge, and efficient system of mass blood donation and supply that is now considered normal. Blood donation did happen in the early twentieth century, but it was 'on the hoof' and ad hoc: because blood storage did not yet exist, if someone needed blood, then someone else had to be fetched to supply it from their arm. The medical procedure was painful and invasive, as the vein was reached by 'cutting down' with a scalpel. If blood hadn't been routinely paid for, hardly anyone would have considered returning. Often, doctors resorted to public servants to supply blood. The police were popular, and in Evanston, Illinois, the police chief begged for firemen to be asked to give blood instead of his officers, who were asked so often, they were looking anaemic. In New York, blood selling was common enough that hospitals were paying $100 a pint and sellers had organised themselves into a union. Blood for transfusion was available, but it was difficult to get, too perishable when you did get it, and it cost money (George 2019).

[2] NHS Blood and Transplant, Annual Report and Accounts 2019/20, HC 773 SG/2020/102.

A 1941 film released by the Ministry of Information about the history of blood donation tells what happened next.[3] The film stars Percy Lane Oliver playing himself, and the scene shows him answering a black telephone to someone from King's College Hospital. They needed blood, so Percy asked his colleagues to donate, Sister Linstead gave a 'pint of the best', and the modern system of voluntary non-remunerated blood donation was cemented. The reality was bittier: voluntary blood donations had happened for years, with men often giving blood to their wives in childbirth. During the First World War, transfusion gained a foothold, but by its end, only about 50 servicemen a day were being transfused. The London Blood Transfusion Service, as Oliver's organisation became known, grew slowly but steadily, despite blood still being bought and sold, and despite logistical difficulties. Donor details, including the nature of the arm to be used, were kept on index cards. But barely any private citizen had a home telephone, so how were donors to be called? One option was the police force. 'Station Officers showed themselves ready to help,' wrote Frederick Walter Mills in a 1949 history of the London Service. 'But donors generally did not like being called upon by a policeman, since they found neighbours were disinclined to take a charitable view of the cause of the visit. One donor accepting the kind offices of the police on such an occasion, had his family's embarrassment increased by being returned to his home in the early hours of the morning by a Black Maria.' (Mills 1949).

By the end of the 1930s, even though the Second International Congress on Blood Transfusion had dismissed a voluntary unpaid blood system as 'hopelessly Utopian' (Anon. 1937, 924), blood donation for no reward was a commonly accepted activity. It was not universal, nor even standard across the UK, but it was getting there. But with war coming, even Oliver's thorough index card system was not enough.

In 1938, most of Britain thought war was imminent. Cardboard coffins were manufactured, gardens dug up, railings removed. Preparations were serious. But one necessity was not: at that time, the total blood supply in London, a city of 10 million people, was eight pints of blood. One woman saw this, and thought it wrong. Janet Vaughan was born to privilege but

---

[3] H. M. Nieter, *Blood Transfusion* (1941). https://wellcomecollection.org/works/hb7wfbuz.

FIGURE 6.2 Janet Vaughan portrait (seated on the right). The Medical Mission for Experts on their visit to India: the group portrait includes Sir Weldon Dalrymple-Champneys (front row, centre right) and Henry E. Sigerist (back row, second from right). Photograph after a photograph by Kundan Lal, c. 1950. Wellcome Collection (CC BY 4.0).

not wealth. Her mother was a society beauty; her cousin was Virginia Woolf. Vaughan was thought 'too stupid to be educated' by one governess, but overcame that, and dyslexia, to get a First from Somerville College, Oxford. That is where I studied and where I first encountered Dame Janet Maria Vaughan, as the name of my first-year accommodation building. I had no idea of her accomplishments, and I still think they are not known and lauded as they should be, because Vaughan was a remarkable woman (Figure 6.2).

She wanted to be a doctor, but her family situation – her father recently widowed, she being the stand-in for her mother – did not allow that. Instead, she moved into pathology, and particularly into blood. First, she was fascinated by anaemia, and by why it was treated with arsenic. Inspired by a paper she had read, she borrowed mincers from all her friends in Bloomsbury – including Virginia – and minced raw liver, feeding the extract first to dogs, and then to herself. She survived, the patients got a much better treatment, and Janet Vaughan was given no credit (Vaughan unpublished).

There was no daunting her. She married, happily, and had children, and continued her bloody studies. Then the Spanish Civil War happened, and she paid attention, thinking herself a Communist. She opened her home to Frederic Durán-Jordà, a pioneering Spanish doctor who had set up an innovative and successful blood supply system in Barcelona (Palfreeman 2015). People came to donate, the blood was stored, and then transported to where necessary. It sounds so straightforward now, but this was the first mass disembodiment of blood from arms to bottles, and it was brave, and revolutionary.

Vaughan saw this, and sheltered Durán-Jordà after he sought exile in England, and began to plot. She asked her head of department if she could devise a similar system, and was sent out with £100 to buy syringes and tubing. She summoned her peers and colleagues to her Bloomsbury drawing room after work, where they plotted and planned every detail. The size of the syringe? The type of bottle? They settled on a modified milk bottle. If London were to be bombed, then it needed a system of mass donation and transfusion. Four depots should do it. The blood would be transported in milk bottles in converted ice-cream vans.[4] The plotters wrote a memo to the head of the Emergency Medical Services. Vaughan's superior, Professor Dibble, came to hear of it and called her 'a very naughty little girl' (Vaughan unpublished). But her memo was heeded, and the system came to pass.

On 1 September 1939, Vaughan received a telegram from the Medical Research Council that read, 'Start bleeding'. On the day that war was announced, she was doing just that in the Northwestern Depot at Slough, a site she had chosen partly because it had a bar. They stopped to listen, the doctors and nurses, then they went back to their bleeding.

How brave they were, the donors, mostly women who wanted to do as much good as they could for the war effort, but also the volunteer drivers, who had to take clinking bottles of blood through streets that had no lights, driving ice-cream vans with no lights, over rubble and under bombing. One of these drivers was Lady Dunstan, who reported for work wearing a hat and always wore pearls. The drivers were so skilled that sometimes they arrived at a hospital with their blood before the casualties who needed it.

---

[4] 1939. Emergency Blood Transfusion Scheme for London and the Home Counties. Now held at the Wellcome Collection.

I have not found records for the Northwestern Supply Depot, but the South West London depot recorded that in 1940 it had distributed 9,410 bottles of blood, and in 1945, 22,397.[5] Newspapers regularly published tales of blood comradeship, which may even have been true, such as Miss M. Lee of Beverley's pint that was flown overseas three days later and transferred to Sergeant Howells, wounded by a mine, and now lacking a leg. Howells survived and Miss Lee was honoured by the Driffield Times for her selflessness[6].

The Army Blood Transfusion Service mirrored the civilian one (Figure 6.3). By the end of the war, the sight of a medic on a frontline with a blood bag hanging from his rifle was ordinary. The remarkable ability of blood from a stranger to heal, revitalise, and restore was no longer questioned. The London Blood Service, the four depots, and the Army Blood Transfusion Service combined into a National Blood Transfusion Service, now NHS Blood and Transplant (NHSBT), which pre-dated the National Health Service by a year. A report by the Medical Research Council after the war honoured the efforts of Dame Janet Maria Vaughan, without whom the story of our blood supply would be very different, with this: 'From the time of the Munich crisis in 1938, the question of blood during wartime had been much in the minds of medical men' (Medical Research Council 1947). And one very naughty little girl.

Villain 2: *Hirudo medicinalis*

It is an annelid worm with 10 stomachs, 32 brains, nine pairs of testicles, and several hundred teeth that leave a bite mark in the shape of a CND symbol or a Mercedes-Benz logo, depending on your preferences. It lives in freshwater, is generously hermaphrodite, can live off one meal for a year, and that meal is blood, and it may be yours.

The thesaurus makes its views of leeches clear: category 'bane', subcategory 'troublemaker', along with parasite, threadworm, and tapeworm. Adolf Hitler loved to call Jews leeches. We still talk of leeching off someone.

---

[5] 1945. South West London Blood Supply Depot (memorandum). London: Ministry of Health.

[6] 1945. 'Beverley Women's Blood Saves Lives in Europe'. *Driffield Times*, 27 January.

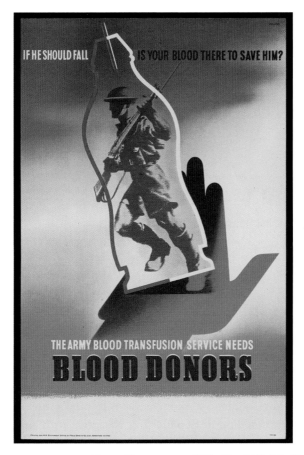

FIGURE 6.3 Abram Games poster 'If He Should Fall Is Your Blood There to Save Him'.

The poor leech has done nothing to deserve this. It may be a parasite, but it is one of the more polite ones, only drinking its fill then dropping off, courteously. Yet it triggers our disgust mechanism, the one that is unleashed by slime (although leeches are not slimy), or bugs, or bile, or, for the phlebo-phobic, blood.

I do not particularly care to handle a leech, but when I was confronted with an Asian buffalo leech – so called because it's not fussy about hairy bovines – I thought it was impolite to refuse. My face wrinkled, I can only remember that it felt cool and alive, and I handed it back as quickly as possible (Figure 6.4).

FIGURE 6.4 Hand holding a leech. Credit: Rose George.

The Asian buffalo leech lived not in a tropical pond, nor on Humphrey Bogart's legs, but just off the M4 motorway in South Wales, at the leech production facility called Biopharm, the only such business in the UK, and one of a few in the world. Leeches are produced because, although they seem mediaeval and barbaric, they are still used in highly advanced microsurgery.

They also have to be produced because human treatment of the medicinal leech throughout history has been so disastrous, it is an endangered species.

For much of human history, humans have treated other humans by removing blood, not adding it. Bloodletting was done with anything that worked, from sharpened reeds to fleams to lancets (hence the name of the eminent medical journal), and for most conditions, including blood loss. On the basis, at least for part of medical history, of the humoral theory, bloodletting was thought to balance the body of excess. The other humours – yellow bile, black bile, and white phlegm – had obvious exit points. 'The body had three doors through which it could evacuate nocive matter', wrote the medical historian Herman Glasscheib (1964), 'through the skin in the form of sweat, through the kidneys as urine and through the bowels as feces. But since there were four juices there must also be four exits. The doctors invented this fourth door in the shape of bloodletting.'

By the end of the nineteenth century people still believed so powerfully in the force of bloodletting that they 'were in the habit of coming to be bled at their own request, just as they now apply to have their teeth drawn' (Thomas 2014). Bleeding was done by opening veins with lancets and knives, but a leech was kinder, sucking its blood from the capillaries, not veins, and providing its own natural anaesthetic. The bite of a medicinal leech – *Hirudo medicinalis* – should feel like a vague sensation rather than a nip or the 'sharp scratch' that favourite but inaccurate warning of phlebotomists and nurses. The leech is in many ways a simple animal, but its anaesthetic and anticoagulant have yet to be bettered by science. Only eight compounds in leech saliva have been identified, and there are probably hundreds that are useful. One, discovered by John Berry Haycraft in 1884, was later distilled into what became known as hirudin, which is vastly more efficient as an anticoagulant than man-made heparin, the next best blood thinner (Kirk and Pemberton 2013). Another is a potent inhibitor of collagen-mediated platelet adhesion and activation. This was isolated after researchers noticed a strange thing: the leech bite made blood flow for hours, but not because of the hirudin (Harsfalvi et al. 1995). Not only is the leech a medicinal treasure chest, but its Mercedes-Benz bite is spectacularly efficient, the tripartite shape much less damaging than a scalpel incision, which can damage the tissue.

The ill fate of the medicinal leech in its natural habitat can be mostly laid at the door of François-Joseph-Victor Broussais, a fashionable nineteenth-century Paris physician who believed all illness arose from 'phlegmasies', an inflammation of the guts, and that the leech was the answer (Broussais 1834). Leech-mania began: at its height, people could be pre-treated with leeches without even being diagnosed, like a leech prophylactic. Broussais prescribed 60 leeches at a time for a grown man, and two to three for a toddler. At the beginning of the century, France had thriving native leech populations. They were gathered by the poor, often young women, who waded into ponds to be bitten, often until they fainted (Figure 6.5). The French called it 'blood fishing'.

> You would suddenly see a young woman soften and sway, as if she were drunk or dizzy, sometimes slump into the pond, her legs in the mud but her head in the clouds. Her companions knew what this flagging meant: a weakness caused by the insatiable vampirism of leeches. So, they would quickly get the stunned girl out of the mud to free her from the slimy parasites.
>
> (Hubert-Pellier 2007)

FIGURE 6.5 Leech finders. *Three Women Wading in a Stream Gathering Leeches.* Coloured aquatint by R. Havell, 1814, after G. Walker. Wellcome Collection (public domain mark).

By 1833, the mania for leeching had grown so huge, France had to import 41.6 million leeches. The popularity of the annelid worm continued, even with notable casualties. Emperor Alexander of Russia died in 1825 after being leeched on his head. Lord Byron, the adventuring poet, submitted to leeches that withdrew two pounds of his blood, when he fell ill in Missolonghi. 'I perceived', wrote one of his doctors, 'that his Lordship had a very great aversion to bloodletting.' 'Come', said Byron toward his end, 'you are, I see, a damned set of butchers. Take away as much blood as you will but have done with it' (Mills 1998). They did, and he died.

Gratitude to leeches was scarce. Lord Thomas Erskine, an eminent politician, was a rarity, because he credited two leeches with saving his life, named them Home and Cline after two Victorian surgeons, and made them pets (Moore 1983). Dr George Merryweather, a physician in Whitby in Yorkshire, made leeches briefly famous for something other than blood loss by inventing the Tempest Prognosticator, a barometer of sorts containing twelve jars, each holding a leech (Merryweather 1851). Merryweather was convinced that leeches could sense storms, and he was not unique in this belief. The eighteenth-century poet William Cowper wrote to his cousin of a storm and of 'a leech in a bottle that foretells all these prodigies and convulsions of nature. ... No change of weather surprises him, and that in point of the earliest and most accurate intelligence, he is worth all the barometers in the world.' (Merryweather 1851).

The Tempest Prognosticator was designed with great care. The jars were glass so that each of what Merryweather called his 'jury of philosophical counsellors' could see its fellows, and not be lonely. But the mechanical barometer required no feeding, it was easier to manage, and it required no suppression of disgust. So, it became the default, and Dr George Merryweather became a footnote in leech history.

By the twentieth century, bloodletting and leeching had fallen from favour with the rise of better surgery and medicine and germ theory. But Biopharm exists because leeches are neither ancient history nor barbaric. Instead, they can often be found in hospital pharmacies, after being rehabilitated in the late twentieth century by a Boston surgeon desperate to save the reattached ear of a young boy called Guy Condelli (Keith et al. 1987). Advanced as we are, we still cannot better the leech for its ability to disentangle clogged blood vessels, to get the blood flowing (Haney 1985). Thirty years after Guy Condelli's operation, the leech occupies a peculiar place in modern life. To the general public, it is simply disgusting. They think leeching is 'evil quackery', says an employee at Biopharm. But when the surgeon Iain Whitaker did a telephone survey of all the 62 plastic surgery units across the UK in 2002, 80 per cent of the 50 that replied had used leeches postoperatively within the last five years. Three units had used leeches more than 16 times a year; 15 had used them up to five times (Whittaker et al. 2004). Leeches are judged to be effective in the salvage of various essential body parts such as fingers, ears, nipples, nasal tips, and penises.

Advanced as we are, surely, we now treat leeches better? They are, after all. advanced medical devices, invaluable. No. The modern leech, once it leaves Biopharm, is a single-use sterile device. Once it has done its job, by saving an ear or a breast or a lip, and although its natural lifespan is 27 years, this villainous parasite is placed in a tub of alcohol and exploded.

Hero 2: Arunchalam Muruganantham

Here again I will swerve away from my hero and start with a villain, and the villain is menstrual blood. This wasn't always the case. In early history, a woman who bled and survived was seen as powerful, even if her powers were clearly imagined by a man. In his multi-volume *Natural History*, the Roman author Gaius Plinius Secondus thought menstrual blood to be both magical and horrifying. He couldn't think, he wrote, of 'anything which is more productive of more marvellous effects than the menstrual discharge'.

On the approach of a menstruating woman, he wrote, nature would cringe and submit. 'Seeds which are touched by her become sterile, grafts wither away, garden plants are parched up, and the fruit will fall from the tree beneath which she sits.' Her look, also, is formidable, because it will 'dim the brightness of mirrors, blunt the edge of steel and take away the polish from ivory'. She can kill a swarm of bees, turn iron and brass rusty. She can scare away hailstorms and lightning, as long as she is both bleeding and naked. At sea, she doesn't even need to bleed: a storm will flee before the sight of her unclothed body. What a useful creature. Farmers must employ their menstruating wives with great joy, because 'if a woman strips herself naked while she is menstruating, and walks round a field of wheat, the caterpillars, worms, beetles and other vermin will fall from off the ears of corn' (Bostock and Riley 1855).

It is telling that many of these menstrual powers came to be associated with witchcraft. Religion took up the theme of our filthy blood with gusto. By the time of the Old Testament, the evil of menstruation was firm enough to be used as an analogy: the book of Isaiah urged the observant to cast away their sinful silver and gold idols as they would a menstruous cloth. Whoever wrote Leviticus was more straightforward. After pronouncing purity rules around leprosy, he moved on to sperm and menstrual blood. 'When a woman has a discharge, if her discharge in her body is blood, she shall continue in her menstrual impurity for seven days; and whoever touches her shall be unclean until evening.'[7] In the thirteenth century, the Jewish scholar Nahmanides also judged the

---

[7] Leviticus 15:19, New American Standard Version.

menstrual woman and found her wanting. 'The dust on which she walks is impure like the dust defiled by the bones of the dead' (Cooper 2016). Most religions agree that a menstruating woman should stay away from God or holy books and places, and they are emphatic about cleansing.

Of course, women like to be clean after menstruating. Without this desire, the sanitary hygiene industry would not have spent two centuries emphasising that their products could make us 'fragrant', because we smell. Of course, women probably like time off from kitchen and marital duties. But I'm suspicious of ritual purity rules. If dirt is matter out of place, then maintaining purity is a matter of putting people in their place. Imaginary dirt is such an effective weapon of limitation. See India's untouchables, imprisoned in filthy jobs – tanning, body removal, latrine emptying – because they are judged filthy. See the most powerful school-yard taunts of disadvantaged children: they are dirty, they reek, they are inferior. See 'you smell', the hardest schoolyard insult to protest. Such a system is an imaginative phenomenon, wrote Virginia Smith in *Clean*, 'that rationality finds so strange – that ritual purity and impurity laws do not refer to observable cleanliness or dirtiness, but to a classified purity status'. You can touch something and not be dirty, but you are unclean. You can bathe in the shit-filled Ganges and be filthy, but you are clean. Mary Douglas once wrote that, to understand purity rules, you have to ask whom they exclude. 'The only thing that is universalistic about purity is the temptation to use it as a weapon' (Douglas 1966; Smith 2007)

Coimbatore is a small town in south India. This is a place where people wear white for the heat, and bananas taste like toffee. I went to Coimbatore to meet a superstar, so when I arrived at his house on a small dusty street, it was a surprise. There he was, waiting outside his house for his guest, though this was a man who has met presidents and won national prizes and (after we met) became the subject of a documentary and a Bollywood film.[8] He led me upstairs to his humble flat, three rooms, one chair, and introduced me to his wife and young daughter. His English was sometimes difficult to understand, as he was self-taught, in English as in everything else, as he had left school at 14 when his father died. But we managed. And over lunch, a delicious vegetable thali served

---

[8] A. Virmani, *Menstrual Man* (2013); and R. Balki, *Pad Man* (2018).

on the floor on banana leaves, I heard the story of Arunchalam Muruganantham, usually shortened to Muruga, but now usually known as Menstrual Man (Venema 2014).

Muruga worked as a machine shop labourer after he had to leave school. He had an arranged marriage to Shanti, and they fell in love. One day, his wife passed him and hid something behind her back. He thought she was teasing and tried to grab it. Finally, she showed him: she was carrying her used menstrual rags to be washed.

Globally, cloth is the most common method that women and girls use to absorb menstrual blood. Although the multi-billion feminine hygiene industry would like us always to wear its commercial products, there is nothing wrong with using cloth. In a way, it's better: it can be reused, and doesn't contain plastic. But cloth is only a good option if it can be cleaned and dried hygienically. In somewhere like south India, that means drying it in the sun, in the open. But the open is out of bounds for most women, who are shackled by the taboo of menstruation and blood. This taboo is not a developing world issue: all women and girls grow up with a surround-sound of messaging. Hide, be quiet, don't show, don't smell. It is effective enough that once, after I had given a talk on menstrual taboo to WaterAid, I looked down at my jeans pocket and worried that my lip-salve looked like a tampon. The advertising industry only dared to show a red liquid in a sanitary pad advert in 2019. Before that, women apparently excreted windscreen-washing fluid.

In India, I met girls and women who were certain that they could cause serious harm while on their period, by touching pickles or going to the temple or looking at boys. In Nepal, the government holds a three-day festival called Rishi Panchami, which was set up to allow women and girls to wash away any sins they may have committed while menstruating. It is wildly popular, and not just because it includes a day off work.

So Shanti did like millions of women do, and hid her drying menstrual cloths. Sometimes they are shoved under a bed to dry, or in cupboards, or tucked away into the dark somewhere. None of this is ideal: badly dried cloths can cause gynaecological infections. Shanti was obviously good at hiding, because this was the first time Muruga had seen her menstrual cloths (though he remembered his sisters hiding theirs in the thatch). He asked her why she didn't buy sanitary pads: he knew they were available

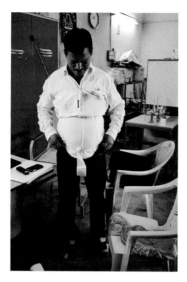

FIGURE 6.6 Arunchalam Muruganantham with sanitary pad. Credit: Rose George.

in the market. She said, because we can only afford sanitary pads or milk, and we need milk more.

At this point Muruga started on his way to being Menstrual Man, because he didn't accept this and carry on. Instead, for the next several years, he tinkered and experimented and tinkered and experimented, to devise a cheap sanitary pad that could be cheap enough for even poor women to buy, and that could be manufactured inexpensively and simply, so that women could sell them too.

It was a long road, because Muruga was starting from profound ignorance. He assumed that sanitary pads were made from cotton, a reasonable assumption, so produced a few simple pads. But how to test them? He looked around and thought that medical students were at ease with the body, so he asked some to test his pads and give him the used ones. Astonishingly, they did. But this wasn't enough for the inventor: he needed to test the pads himself. He made a uterus out of a football, filled it with goat blood, attached it to a tube with a pump, fed another tube into his pants, and went about his business (Figure 6.6).

In the documentary *Menstrual Man*, Muruga describes how he was always turning around, checking for leaks. 'I'd become a woman.'

The townspeople began to talk. He was a vampire, or a monster, or an adulterer, hanging around young women not to examine their menstrual pads but for adultery, which was obviously much worse. Shanti left him. His mother almost disowned him. When he heard rumours that he was to be chained to a tree, he fled. Deprived of his family, he was undeterred. By now he suspected that the big companies were using something more complex than cotton. Finally, by pretending to be a manufacturer, he was given samples of cellulose, and he understood. This was the absorbency he needed.

Years later, he made a machine of 243 parts that needed very little electricity, relied largely on foot pedals, and could be operated by anyone. In 2006, it won first prize at the Indian Institute of Technology Awards. He was judged one of *Time* magazine's top 100 people, alongside Edward Snowden. At his flat, he showed me a cupboard where he keeps his awards, and it is crammed full. I noticed that in every photograph he was wearing glasses, but he wasn't wearing them now. He said, 'Once I was being picked up at an airport, and I went over, and the man said, "No, no, we are looking for a VIP." So now I wear glasses. I don't have any eye problem, they are just to give some lift so the [drivers] will take me. My VIP glasses.'

The man with the fake VIP glasses has seen his machine thrive. There are thousands in India, and plenty in other countries too. Muruga became famous, and his wife came home.

I love the story of Muruga, and he is heroic. But there are countless women who have been working on improving sanitary hygiene who have not become VIPs or the subjects of films or the friends of presidents. And in a way the last thing India needs is more sanitary pads, as it already can't manage the millions that are already used, that clog small sewers or are thrown into fields, that are rarely properly disposed of.

Still, when menstruation is still considered so filthy, the Prime Minister Gordon Brown couldn't bring himself to say 'tampon' at the dispatch box (McBride 2013), and medical men up to the 1970s were discussing whether women emitted a 'menotoxin' that could wilt flowers and make dough rise 22 per cent less (Pickles 1974): even with the greater visibility of periods, there is still progress to be made and we still need all the heroes we can get. Because we are still too far from the alternative reality perfectly described by Gloria Steinem (1978) in her essay 'If men could menstruate'. In this other world:

Men would brag about how long and how much.

Young boys would talk about it as the envied beginning of manhood. Gifts, religious ceremonies, family dinners, and stag parties would mark the day.

To prevent monthly work loss among the powerful, Congress would fund a National Institute of Dysmenorrhea.

Sanitary supplies would be federally funded and free. Of course, some men would still pay for the prestige of such commercial brands as Paul Newman Tampons, Muhammad Ali's Rope-a-Dope Pads, John Wayne Maxi Pads, and Joe Namath Jock Shields 'For Those Light Bachelor Days'.

References

Anon. (1937) 'Blood Transfusion Congress in Paris'. *British Medical Journal* 2, 923–924.

Blundell, J. (1834) *The Principles and Practice of Obstetricy, as at Present Taught.* London: E. Cox.

Bostock, J., and Riley, H. T. (transl.) (1855) *The Natural History of Pliny.* London: H. G. Bohn.

Broussais, F. O.-J.-V. (1834) *Cours de pathologie et de thérapeutique générales.* Paris: J. B. Baillière.

Cooper, C. (2016) *Blood: A Very Short Introduction.* Oxford: Oxford University Press.

Coxe, T. (1666) 'An Account of Another Experiment of Transfusion, viz. of Bleeding a Mangy into a Sound Dog.' *Philosophical Transactions of the Royal Society* 2, 451–452.

Denis, J. (1666a) 'A letter concerning a new way of curing sundry diseases by transfusion of blood, written to Monsieur de Montmor, Councellor to the French King, and Master of Requests.' *Philosophical Transactions of the Royal Society* 2, 489–504.

Denis, J. (1666b) 'An Extract of a Letter Written by J. Denis, Doctor of Physick, and Professor of Philosophy and the Mathematicks at Paris, touching a late Cure of an Inveterate Phrensy by the Transfusion of Blood.' *Philosophical Transactions of the Royal Society* 2, 621.

Douglas, M. (1966) *Purity and Danger. An Analysis of the Concepts of Polllution and Taboo.* New York: Routledge.

George, R. (2019) *Nine Pints: A Journey through the Mysterious, Miraculous World of Blood.* London: Granta.

Glasscheib, H. S. (1964) *The March of Medicine: The Emergence and Triumph of Modern Medicine*. New York: Putnam.

Haney, D. Q. (1985) 'Doctors Combine Modern Microsurgery and Ancient Leeching to Save Ear.' Associated Press. https://apnews.com/article/06ff6e32bf915783749bd0473b0a686b.

Harsfalvi, J., Stassen, J. M., Hoylaerts, M. F., Van Houtte, E., Sawyer, R. T. et al. (1995) 'Calin from *Hirudo medicinalis*, an inhibitor of von Willebrand factor binding to collagen under static and flow conditions.' *Blood* 85, 705–711.

Hubert-Pellier, M. (2007) 'La pêche à la sangsue: échos des pratiques pénibles d'un commerce insolite.' *Amis du Vieux Chinon*, 11(1), 39–48.

Keith L., Mutimer, J. C. B., and Upton, J. (1987) 'Microsurgical reattachment of totally amputated ears.' *Plastic and Reconstructive Surgery* 79, 535–541.

Keynes, G. (1922) *Blood Transfusion*. London: Henry Frowde and Hodder & Stoughton.

Kirk, R. G. W., and Pemberton, N. (2013) *Leech*. London: Reaktion Books.

Lower, R. (1665–1666) 'The Method Observed in Transfusing the Bloud out of One Animal into Another.' *Philosophical Transactions of the Royal Society* 1, 353–358.

McBride, D. (2013) *Power Trip: A Decade of Policy, Plots and Spin*. London: Biteback.

Medical Research Council (Great Britain) (1947) *Medical Research in War: Report of the Medical Research Council for the Years 1939–45*. London: HM Stationery Office.

Merryweather, G. (1851) 'An Essay Explanatory of the Tempest Prognosticator in the Building of the Great Exhibition for the Works for the Industry of All Nations'. Read before the Whitby Philosophical Society, 27 February. https://archive.org/stream/b2804163x/b2804163x_djvu.txt.

Mills, A. R. (1998) The last illness of Lord Byron. *Proceedings of the Royal College of Physicians Edinburgh* 28, 76.

Mills, F. W. (1949) 'The London Blood Transfusion Service and the psychology of donors.' In G. Keynes (ed.) *Blood Transfusion*. Bristol: John Wright & Sons, pp. 346–368.

Moore, T. (1983) *The Journal of Thomas Moore*, ed. Wilfred S. Dowden. Newark, DE: University of Delaware Press.

Palfreeman, L. (2015) *Spain Bleeds: The Development of Battlefield Blood Transfusion during the Civil War*. Eastbourne: Sussex Academic Press.

Pepys, S. (1667) *Diary of Samuel Pepys*. London.

Pickles, V. R. (1974) 'Letter'. *Lancet* 303, 1292.

Smith, V. (2007) *Clean: A History of Personal Hygiene and Purity.* New York: Oxford University Press.

Steinem, G. (2019) 'If men could menstruate'. *Women's Reproductive Health* 6(3), 151–152.

Thomas, D. P. (2014) The demise of bloodletting. *Journal of the Royal College of Physicians of Edinburgh* 44(1), 72–77.

Vaughan, J. (unpublished) *Jogging Along, or, A Doctor Writes.* Somerville College, University of Oxford (unpublished manuscript).

Venema, V. (2014) 'The Indian sanitary pad revolutionary'. www.bbc.co .uk/news/magazine-26260978.

Whitaker, I. S., Izadi, D., Oliver, D.W., Monteath, G., and Butler, P. E. (2004) '*Hirudo medicinalis* and the plastic surgeon.' *British Journal of Plastic Surgery* 57, 348–353.

7 Cold Blood: Some Ways by Which Animals Cope with Low Temperatures

STUART EGGINTON

Extreme physiology, the body's capacity to cope with external or internal challenges, can take many forms, but here we are concerned with one of the limits to exercise in otherwise unfavourable conditions. The presence of cold blood, whether through acute seasonal chill or chronic environmental exposure, imposes an additional burden on the heart's ability to pump this viscous fluid around the body, potentially limiting perfusion of working muscle. Also of interest is how the 'business end' of the cardiovascular system, the microcirculation, adapts under these conditions. Here, intimate contact between blood and tissue is achieved by a vast network of tiny vessels (capillaries) that facilitate supply of oxygen and other fuels, as well as removal of waste products. These smallest of blood vessels thus influence both health and disease, although other elements of the cardiovascular system need to work in harmony for optimal performance. This chapter will explore the physiological response to blood that is cooled, in different animals and across different timescales. To briefly illustrate the adaptive capacity and great diversity of responses observed, we will consider some strategies that endotherms ('warm-blooded' mammals) use to cope during winter, and contrast this with adaptations seen in ectotherms ('cold-blooded' fishes) that thrive in the constantly frigid waters around Antarctica.

Background

The following will examine some consequences of cold exposure and what being cold blooded means, not as a descriptive literary device but exploring the physiological response to blood that is actually cooled. Professor Pedley (Chapter 3 in this volume) has discussed the difficulties of perfusing blood, the viscous fluid so essential to life; this becomes much harder to push around our blood vessels when cooled down and so cold exposure represents a particular challenge for all animals. Rose George (Chapter 6 in this volume) has listed a number of important functions of blood, so it should be no surprise that there are many ways to optimise bodily functions when it is cooled. By considering a few examples across a range of animals and timescales, comparing the adaptive capacity of mammals and fishes, I hope to provide a flavour of the tremendous diversity of responses that are possible.

What Is Meant by 'Cold-Blooded'?

Most people think of this in terms of something cool to the touch, especially with reference to our own body temperature, like a gecko running up a boy's shirt or a snake wrapped around somebody's arm (Figure 7.1), but all animals experience thermal challenges over acute, daily, seasonal, and (during speciation) evolutionary timescales.

Thermal biologists tend to group animals into large categories. The first comprises endotherms like ourselves, often termed homeotherms from the Greek, meaning *similar heat*, or inappropriately termed 'warm blooded'. However, if you go out on a winter's day without gloves you will understand the meaning of cold blood, while the garden hedgehog

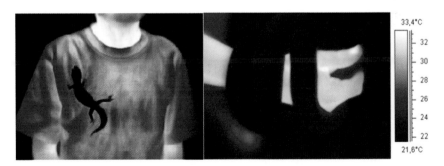

FIGURE 7.1 Contrasting body temperature of some animals and humans. We tend to think of animals that have a lower body temperature than ours, such as a gecko (left) or snake (right), as being somehow inferior, so the term 'cold-blooded' can have a negative context; nevertheless, the majority of animals on the planet come under this category (Figures courtesy of NASA/IPAC under Creative Commons licence BY-SA 3.0).

will be hibernating after dropping its core temperature to just a few degrees above ambient. Endotherms generate endogenous heat by metabolism, which gives a degree of independence from the external environment; hibernators may have really cold blood, but this represents regulated, not passive cooling. The other large group is ectotherms, which used to be called poikilotherms (*varied heat*), that are largely dependent on the external thermal environment. It is also a misnomer to call them 'cold blooded'. TV documentaries show lizards running across the desert lifting up their feet because the sand is too hot to walk on: these are definitely not cold-blooded reptiles. As with any biological definition, there are always exceptions to the rule: the mesotherms, which is a sort of intermediate stage. These animals can generate metabolic heat, but they cannot thermoregulate as well as endotherms, and this can be for a number of reasons. It could be a scaling factor: even a sleeping 10-ton *Tyrannosaurus rex* will be generating a huge amount of metabolic heat that is unlikely to be fully dissipated to the environment, so this is not going to be a cold-blooded animal. It could be anatomical: the skipjack tuna is the fastest fish in the ocean because it has an anatomical heat exchange system that captures the metabolic heat generated by its swimming muscles, and it is a couple of degrees warmer (and therefore faster) than its prey. Actually, *all* animals are heterotherms because there are regional variations in temperature; this differs among types of tissue, over time, between species, and across ecological niches.

Thermoregulation

Body temperature (T_B) is maintained by a number of mechanisms, including behavioural, evaporative, vascular, and metabolic responses; in homeotherms T_B is tightly regulated through homeostasis that provides a relative independence of external factors, whereas in ectotherms T_B is largely dependent on the thermal environment (Figure 7.2(a)). Body temperature normally fluctuates over the day following circadian rhythms; in humans this reaches a minimum in the early morning and a maximum in the evening, with females on average slightly warmer than males (Figure 7.2(b)). Behavioural thermoregulation is also important: a lizard warming itself up on a stone wall in the morning is effectively

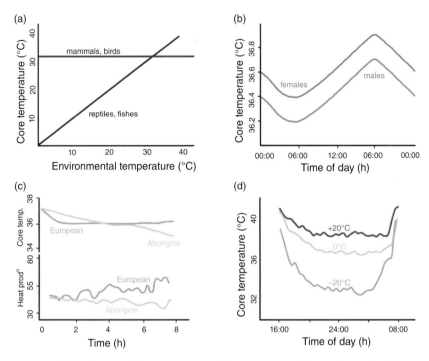

FIGURE 7.2 Core temperature fluctuates over different timescales. (a)
Contrasting mean core temperature (T_C) of endotherms (red line) and ectotherms (blue
line) in idealised response to changes in environmental temperature, illustrating the
relative independence and dependence, respectively. (b) Diurnal variation in human T_C of
around 0.5 °C, being lowest at 4–5 a.m. and highest at 5–6 p.m., first noticed by Carl
Wunderlich (1868); females (brown line) are on average slightly warmer than males
(green line). (c) Thermal responses of Aborigines (yellow lines) and Europeans (blue lines)
during cold-air exposure, sleeping essentially naked at ~5 °C for eight hours; the latter
defended T_C but incurred a metabolic cost in doing so. After Hammel et al. (1959). (d)
Willow tit with a radio telemetry transmitter showing how T_C is adjusted to conserve
energy expenditure in proportion to altered thermal gradient across seasons. After
Reinertsen (1982).

transitioning between cold blood and warm blood. Other cycles may be
present due to hormonal (e.g., the rhythm method of contraception),
feeding (due to metabolic cost of digestion), and seasonal (depending on
latitude) factors.

It is important to recognise that we are endothermic but not isother-
mic, i.e., not all of our body is at the same temperature. Heat exchange
occurs by physical mechanisms (radiation, convection, conduction,

evaporation), and we can modulate, but not completely exclude, these influences. In a cold environment we will tend to vasoconstrict our periphery in order to conserve heat in the core. In warm environments, we will flood the periphery with blood by vasodilatation in order to dissipate that metabolic heat and avoid overheating: while we regulate core temperature, shell temperature can vary, and the shell can be perfused by relatively cold blood.

Heterothermy

Is there any evidence of adaptation to different environmental temperatures? In deserts it gets very hot during the day, but at night it gets very cold. A study in the 1950s of Aboriginal Australians and Kalahari San noted that, around evening campfires, those of European descent would be desperately shivering to keep warm, whereas their companions would be sleeping quite soundly without shivering. By checking core temperatures with a rectal thermometer, it was evident that the former defended their core temperature, but at a cost of raised metabolic rate, while the latter allowed their core temperature to decrease, thus reducing metabolic cost (Figure 7.2(c)). Shivering thermogenesis (uncoordinated muscle contractions used to keep warm) can lead to a five-fold increase in heat production, so is this an adaptive energy-saving response to cold exposure?

From a survival perspective, such a response would have an adaptive value for animals, i.e., represent a selection pressure, especially for small mammals and birds with a high surface-to-volume ratio where it is easy to dissipate heat, but much more difficult to conserve it: if core temperature drops, it costs a lot to raise it again. For example, in a small bird, circadian rhythms of body temperature are increased by environmental challenges (Figure 7.2(d)): even in balmy summer evenings (+20 °C) there is a small overnight drop in T_B; in autumn when the nights are chilly (0 °C) the drop in temperature is more pronounced; the depths of winter provide a really challenging environment (−20 °C), and where food is also scarce a considerable energy saving can be made by allowing T_B to drop during the night. This happens over a daily cycle, but a similar sort of hypometabolism is observed in hibernators: this may last for a few weeks in small mammals and for some months in larger mammals.

But we must be careful to avoid generalisations. This lecture was delivered in the week of the 150th anniversary of the publication of *The Descent of Man* by Charles Darwin, after whom the College is named. During his trip on *The Beagle* he commented on the remarkable resilience of the Alacalufe tribe of Tierra del Fuego, a very inhospitable part of South America where it is always wet and cold. Again, the indigenous population would sleep in rough shelters with little insulation, and again apparently without shivering. A repeat experiment in the 1960s confirmed they were not shivering, but in this case metabolic rate was high and thus defending their core temperature. This appears to be a similar mechanism to that underlying the cold tolerance of high-latitude populations such as Inuit and Saami, who eat a high-protein diet, the high calorific content allowing them to stoke the 'fire of life'.

Functioning in the Cold

Physiological limits to sustained activity in the cold depend on the individuals concerned, the types of animals, and the various challenges involved. Whether our team is sampling on top (Figure 7.3(a)) or under (Figure 7.3(b)) the frozen sea around Antarctica, we, endotherms, rely on good insulation to cope with extreme environments. Other animals also have good insulation, which we often emulate, but still face challenges with peripheral circulation when perfused with cold blood. For example, the thick blubber layer of seals protects core temperature, establishing a steep thermal gradient from the cold skin to warm muscles, but superb insulation impinges upon tissue repair – males haul themselves out onto sea ice to heal wounds from territorial disputes and allow those regions to warm up. Likewise, special organs (eyes, nose) need to operate at a higher temperature than the surrounding water as cold blood is not conducive to sensory acuity, so there needs to be a compromise of losing heat from areas perfused with warm blood (Figure 7.3(c)).

Other anatomical adaptations to conserve heat are exemplified by the often-asked question 'Why don't penguins melt snow?' The answer came from a famous experiment conducted by Knut Schmidt-Nielsen, one of the founders of comparative physiology, working on ducks, but the principle is broadly applicable (Figure 7.3(d)). As endotherms experience

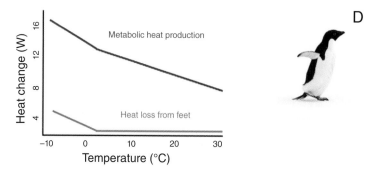

FIGURE 7.3 Humans and animals in extreme cold. (a) Humans emulate endothermic animal insulation to work in the cold. The author travelling over the frozen sea on skidoo to go ice fishing off Cape Evans, Antarctica; many layers of clothing are required to maintain body temperature (T_B). (b) Retrieving telemetered fish from off Adelaide Island, Antarctica, diving with a dry suit to minimise cold exposure. Photo courtesy of Dr Hamish Campbell, Charles Darwin University. (c) Thermal images showing the efficacy of insulating blubber (blue = cold) around the body of a Weddell seal compared with eyes and nares (warm = red), where tissue temperature is raised. Courtesy of Dr Terrie M. Williams, UCSC. (d) Conservation of body heat in birds by a specialised anatomical arrangement of blood vessels called a rete mirabile – Latin for 'wonderful net' – that exchanges heat from arteries carrying warm blood to its feet with veins carrying cold blood from the feet; this minimises heat loss for the feet (blue line) and resultant metabolic cost (red line) of maintaining a high T_B. After Kilgore and Schmidt-Nielsen (1975).

cold they increase metabolic heat production to defend core temperature, so you might expect a dissipation of heat to the environment from naked feet of birds. In fact, the heat loss is minimal until just below 0 °C, at which it increases a little bit. This is possible due to an anatomical adaptation called a *rete mirabile*, found in lots of birds, that acts as a counter-current heat exchanger. If warm blood from the core went to the feet directly then it would lose heat to the environment, but instead venous blood is cooled by walking on snow, then cools the arriving blood, while warm blood from arteries warms up the blood returning to the core; this avoids both heat loss from the feet and chilling the core. By reverse plumbing, animals can walk across deserts without overheating their heads in the midday sun, having a heat exchanger to cool the brain.

It is important to understand that temperature will affect biology at all levels; it is pervasive, and will have a direct or indirect effect on every biological function. So we need to think from a holistic view about how we respond to cold to preserve activity: it may be a change in ventilation and blood oxygenation, a change in the capacity to pump blood around the body or of the blood vessels to accommodate this cold blood, or a change in the ability of the microcirculation (especially the capillary bed) to deliver oxygen by diffusion, or in the ability of the mitochondria within cells to generate adenosine triphosphate (ATP) that determines muscle performance. We will consider just a few of those responses.

Cold Paws Require Large Hearts

In endotherms there are broadly two responses to winter temperatures, which can be described as either a 'capacity adaptation' (a strategy that allows animals to maintain activity, as seen in rats running through icy streets) or a 'resistance adaptation' (whereby animals adopt accommodation measures to minimise the debilitating effects of cold on physiology, e.g., hamsters undertaking bouts of hibernation). On cold exposure most animals will show a reduced cardiac output, partly because of increased blood viscosity, which will impair oxygen transport to tissue and hence reduce motor activity. Both survival strategies require a range of compensatory responses, a common one is cardiac hypertrophy (an increase in heart mass). Summer-acclimatised animals (22 °C) were placed in an environmental chamber with a slowly decreasing temperature (to 4 °C) and shortening photoperiod to mimic the onset of winter for eight weeks. This produced a 30 per cent increase in the relative cardiac mass (percentage of body mass occupied by the heart) in rats and a 20 per cent increase in hamsters (Egginton et al. 2001). To put that in context, growth of the heart is two to three times greater in these animals in accommodating cold than humans achieve in developing athletic capacity.

Having a larger pump is potentially advantageous, but how well does that work? Using acute cooling of core temperature, we can examine changes in cardiac activity over a safe range of temperatures. We see a high heart rate at 37 °C, slow heart rate at 25 °C, returning to a high heart rate when rewarmed to 37 °C, at which electrical connectivity in the heart (via the electrocardiogram) is restored (Figure 7.4(a)). As the temperature is lowered, hamsters show a regular decrease in heart rate, a vagal-induced bradycardia (reduced heart rate); the parasympathetic system is slowing the heart down in proportion to the decrease in core temperature (and hence metabolic demand that the heart is supporting). Rats begin with a decrease and then an elevation of heart rate, at this point the animal starts to defend its core temperature by invoking shivering thermogenesis; this is energetically costly and so the heart rate goes up. Once shivering ceases, the drop in heart rate parallels that of hamsters. In both of these animals heart rate is reduced by 40–50 per cent with acute cooling, but the blood pressure decreases by only 5–10 per cent. This suggests the rest of the cardiovascular system is accommodating bradycardia in order to preserve blood pressure, which is essential for perfusion of tissues. Interestingly, the coupling between heart rate and ventilation rate is maintained over this temperature range. This demonstrates that neural cardiorespiratory coupling remains intact, although at lower temperatures it starts to fail, ultimately leading to circulatory collapse (Hauton et al. 2011).

On chronic cold exposure (leading to acclimatisation) a partial compensation improves the metabolic capacity of the rat heart, but when using isolated hearts (a Langendorff preparation, used since 1895 to study cardiac pharmacology; see Bell et al. (2011)) in the hamster, something strange happens. We know the enlarged hearts in cold hamsters grow new capillaries (angiogenesis); all else being equal, this should lead to a better-performing heart. But despite this, the power output of the cold-acclimatised hamster heart is less than the output from euthermic (normal environmental temperature and T_B) controls. This apparent paradox reflects altered autonomic neural regulation in response to differing physiological challenges such as changes in blood pressure.

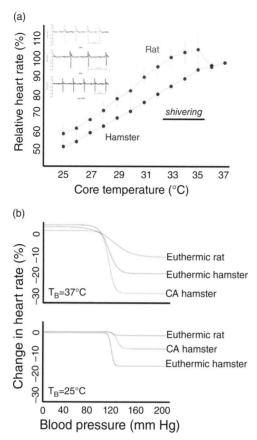

FIGURE 7.4 Changes in heart rate on cold exposure. (a) Experimental lowering of mean core temperature (T_C) induces a change in heart rate for hamsters (blue circles) and rats (red circles) when cooled to 25 °C then rewarmed to 37 °C, without compromising cardiac electrical activity (seen by the ECG, insert). After Hauton et al. (2011). (b) Baroreceptor reflex curves for normal (euthermic) and cold-acclimatised (CA) hamsters, demonstrating the response to changes in blood pressure by means of modulating heart rate at different temperatures, in comparison with rats. From Egginton et al. (2013), with permission.

The baroreceptor reflex curve (Figure 7.4(b)) of rats shows they can accommodate a normal dynamic range of adaptive responses that we expect in a mammal at 37 °C, which essentially disappears at 25 °C: this represents a real danger zone where autonomic regulation is lost, hence the need to defend T_B. In contrast, hamsters accommodate blood pressure fluctuation well at 37 °C and maintain this regulatory capacity even on acute cooling: they are hardwired to cope with the cold. Paradoxically, cold-acclimatised hamsters become hypersensitive at warm temperatures and desensitised at low temperatures, suggesting that winter survival may compromise elements of cardiovascular function to preserve those more critical for subsequent rewarming (Egginton et al. 2013).

These data and others concerning the reactivity of blood vessels (e.g., capacity for vasoconstriction and vasodilatation) suggest that a cold-acclimatised hamster is adopting a compromise position. To enter hibernation, where they have to deal with cold blood for a long time, these animals decrease temperature beyond the 'magic zone' around 20 °C, where a conflict arises in autonomic regulation, between the parasympathetic break on heart rate and the sympathetic drive for cardio-acceleration. At this point there can be a paradoxical rise in stimulation of the heart, which is maladaptive and leads to fibrillation: entering hibernation is dangerous, and of course turning the corner and raising temperature during arousal is also dangerous. The heart has to cope with performing at very low temperatures and then, over a very short period of time, perform at high temperatures. These animals will arouse from hibernation from 4 to 37 °C, with shivering thermogenesis generating all the heat required, in a matter of 60–90 minutes; this is a huge physiological challenge.

Cardiac Changes Are Complemented by Vascular Reactivity

If you put your hand in ice-cold water, then you get vasoconstriction (the fingers turn a lighter colour), namely blood vessels are actively narrowed to reduce perfusion of the shell and to preserve the temperature of the core. But after a while you will get a warming of the fingers – paradoxical cold-induced vasodilatation – which makes adaptive sense because, if you continue to prevent blood flow, tissue will become damaged by hypoxia, frostbite will develop, and eventually digits will be lost. This peripheral survival strategy tends to be centrally mediated, as innervation of the vascular smooth muscle regulates this vascular tone. That mechanism may be activated excessively in Reynaud's disease, sometimes called 'white finger' or 'wax finger'.

Are there any adaptive possibilities here? The problem with decreasing heat loss by cutaneous vasoconstriction is that it comes at the cost of manual dexterity. I know this from personal experience when doing fine surgical manipulation in Antarctic aquaria at sub-zero temperatures; on first arrival I can manage a couple of minutes before losing the capacity to work. I need to warm my hands up, which can be painful. After a few weeks of repeated cold exposure, manual dexterity is possible over a matter of 10–15 minutes. We can see this adaptive response in other populations as well. Korean pearl divers (Ama) used to dive in water at around 10 °C without insulated wetsuits, so they also experienced repeat cold exposure. These women, relative to their non-diving compatriots, had a blunted vasoconstrictor response that maintained blood flow to the periphery longer than normally would be expected, thereby prolonging activity in the cold.

Any neural regulation of vasoreactivity must be mediated by a chemical neurotransmitter, with transduction of the signal from nerves to vascular smooth muscle involving noradrenaline, so peripheral failure as well as a central failure may be implicated. We recruited our cold hamsters again to explore this possibility, using the cheek pouch viewed under an intravital microscope to examine living tissue. When noradrenaline is added topically to terminal arterioles, the 'resistance vessels' regulating microvascular perfusion constrict and shut off blood flow to the downstream capillaries. If the animal is acutely cooled, and vessels are perfused by cold blood, then vasoreactivity to noradrenaline is lost: the ability to redistribute blood around the vascular bed according to need is no longer present. However, noradrenaline vasoreactivity is restored after cold acclimatisation, so there is also a peripheral adaptive response to cold, and we have some idea of the mechanisms involved. (This result is from the author's unpublished data.)

Capillary Supply Adjusts to the Cold

The microcirculation acts as an interface between blood delivery and tissue consumption of oxygen and other nutrients, as well as achieving removal of metabolic waste products, by the process of diffusion. This requires a multitude of tiny vessels, both to minimise transport distances involved and to maximise exchange surface area. Arterioles branch to support many capillaries, a network filled with red blood cells (erythrocytes), which drain into venules. The orientation of capillaries is largely along the axis of muscle fibres, but where oxygen demand is high this pathway is more tortuous (Figure 7.5(a)). There is a lot of variation: regions of high or low capillary density; some capillaries with a high proportion of red cells (high haematocrit), others with low haematocrit; and capillaries with high blood flow or with very low blood flow. These characteristics change with time, depending on tissue demand. Importantly, all of those variations will influence delivery of oxygen to working muscle – and ultimately our performance – so regulation of oxygen supply and demand is affected by very-small-scale adjustments in function of microscopic structures.

Let us take a look at what happens to the capillary bed on cold acclimatisation. We examined biopsies of hindlimb muscles of hamsters used during walking in cold (ambient temperature 4 °C) vs. euthermic (room temperature) conditions – paralleling winter vs. summer acclimatisation. Capillary growth (or 'angiogenesis') was evident on cold exposure – in response to slowing of diffusion in the cold, the distance over which diffusion occurs has been reduced; oxygen exchange has been improved, with more sources of oxygen at the microscopic level. In addition, the oxidative capacity of the muscle also increased on cold acclimatisation, along with a shift towards slower contracting (Type 1) fibres. These are the fibre types that are recruited first in the motor programme when using small movements, adjustments in posture and so on, that you may expect in a cosy burrow. In contrast, the fast (Type 2) muscles that are used for ballistic activity, no longer required for escape responses, are not well preserved; hamsters effectively accommodate the onset of winter. There was no effect of cold acclimatisation on capillary supply or fibre size in rats, these are nicely adjusted and co-regulated to avoid the detrimental effects of cold exposure: despite exposure to cold air, they display a normal phenotype, maintaining their core temperature by feeding more (hyperphagia) and by increased activity (hyperkinesia), so perhaps that is not surprising. Hamsters, however, undertake disuse atrophy; they are getting ready for a long winter sleep, so they are not active, and consequently muscle wasting generally reduces fibre size.

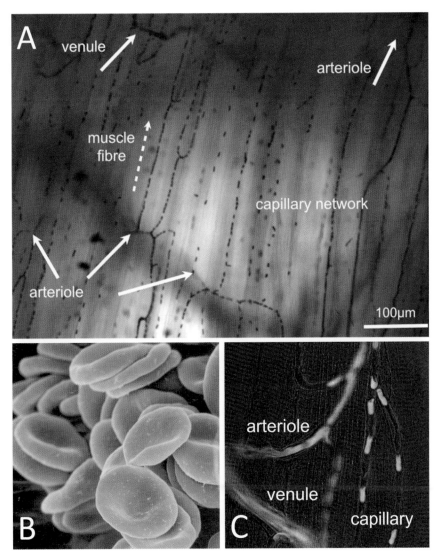

FIGURE 7.5 The capillary system and erythrocytes. (a) The capillary bed in rat skeletal muscle – there is variation in spacing, haematocrit, and perfusion among capillaries (×10 intravital microscope image). Photo courtesy of Professor Chris Ellis. (b) Human erythrocytes, showing the discoid shape normally seen in suspension or when flowing in large blood vessels (scanning electron microscope image). CC licence, photo credit Tina Carvalho https://images.nigms.nih.gov/). (c) Erythrocytes deform in order to squeeze through capillaries, rat skeletal muscle (×40 intravital microscope image). Photo courtesy of Professor Chris Ellis.

Interestingly, in regions of muscle closest to the centre of the leg (and therefore the warmest) there is capillary growth, while in the outer part of the muscle, which is glycolytic and therefore less reliant on oxygen, there is capillary loss (Egginton et al. 2001). This is really exciting because, although we know there can be different growth processes among organs, this may be the first time both angiogenesis and capillary rarefaction have been observed in the same tissue. Again, this emphasises that the response to the challenge of cold blood is not global, but very specific and very local.

The final part of the cardiovascular system that influences oxygen delivery to working muscle comprises the erythrocytes, deformation of which is key to microvascular perfusion. The usual images in textbooks are of disc-shaped erythrocytes (Figure 7.5(b)), which are normally like this only in suspension or when flowing in large blood vessels. When passing through capillaries, they turn into tubes (Figure 7.5(c)). Since they are slightly bigger than the capillary lumen, they have to squeeze through; in doing so they maximise the exchange of oxygen bound to haemoglobin to match demand for oxygen from mitochondria in muscle. Capillaries are similar irrespective of the vertebrate examined: fishes, birds, mice, and elephants all have essentially the same size of capillary. However, the biophysical nature of membranes is temperature-dependent; they become stiffer and the erythrocyte less deformable on cooling, so that when blood is cooled there are likely to be additional problems with microvascular perfusion.

Fishes and birds have an additional problem, they have nucleated erythrocytes, which are bigger and less deformable than mammalian erythrocytes; so fishes potentially face significant problems with microvascular perfusion, and should be prime candidates in which to observe adaptive responses. What happens if we cold acclimatise these animals, can we adjust erythrocyte deformability? The answer is no, membrane rigidity is apparently non-adaptive (Nash and Egginton 1993). So, despite the fact that we have a plethora of responses at many different levels of organisation, from the heart and larger blood vessels to the microcirculation, as well as other factors, including innervation pattern and enzyme activity – there comes a biophysical limit to how far adaptation to cold blood can reach. A barrier has been hit at this point, beyond which membrane deformability cannot be adjusted.

Challenges of Cold Blood for Ectotherms

We will now look at the second main type of thermal biology found among vertebrates, the ectotherms, using fish as our focus. Fishes are great subjects to explore potential responses to cold exposure, and uncover some of the mechanisms involved that may be of benefit in clinical and veterinary science. They constitute the largest and most diverse group of vertebrates on earth, with around 30,000 species of bony fishes. For the current topic what is really exciting is that they occupy thermal niches from East African lakes that are warmer than our body temperature, to sub-zero waters around Antarctica. Gills are gas-exchange organs (aquatic 'lungs') with an enormous surface area; this also makes them efficient heat-exchange organs, so the core temperature is usually very close to that of the environment. Fish generally maintain activity across this vast range of water temperature, so they are clearly well able to deal with cold blood.

Humans tend to be very interested in adaptations to high heart rates, high blood pressure, and large hearts because from an anthropocentric perspective these are pro-arrhythmogenic (lead to arrhythmias – disturbed heartbeats), and potentially dangerous. Much less well studied is the propensity of low heart rate and blood pressure to elicit arrhythmias; but most animals on the planet exist perfectly well with lower heart rates, lower blood pressure, and indeed lower body temperatures than we humans. An understanding of the range of adaptive responses seen across endotherms and ectotherms is helpful both in our understanding of how individual species react to environmental challenges and in offering a broader perspective. After all, the vertebrate pattern is very similar, irrespective of species considered: the mechanisms involved in cardiac contractility in a fish are not that dissimilar from those working in apes, albeit requiring, for example, lower calcium levels to elicit contractions, and adopting some variations in proteins involved.

To emphasise some common responses that we see in endotherms and ectotherms, consider the cardiac response to cold. Hearts are smaller in fish than in mammals as this varies with oxygen consumption – the metabolic rate is lower and the requirement for cardiac activity is consequently reduced. But for Crucian carp held at summer temperatures of

25 °C, then acclimatised for a number of months to 5 °C, we see a 30 per cent increase in relative cardiac mass, very similar to that seen in a cold-acclimatised mammal at 4 °C (Young and Egginton 2011). As biological fluids are cooled, their viscosity is increased, so it is hard to perfuse tissue. Hearts work hard to accommodate viscous blood and maintain cardiac output; just as you build skeletal muscle by working out in the gym, these hearts hypertrophy in response. However, fishes that live in the sub-zero temperatures of the Southern Ocean around Antarctica, such as *Notothenia coriiceps*, have to cope with a thermal challenge not just on a seasonal basis but all year round. They are constantly cold. Not surprisingly, their cardiac mass is even higher, some 40 per cent greater than 5 °C-acclimatised carp (Joyce et al. 2018b).

Considering peripheral oxygen transport, and diffusive limits that are overcome by the capillary bed, do we see parallels there as well? Temperate-zone fishes, those that live somewhere between high and low latitudes, have a seasonal cycle of temperature, so we explored this with the help of a friendly fish farmer whose rainbow trout are held in outdoor raceways. They are subject to the vagaries of weather, i.e., changes in photoperiod and temperature. We looked at muscle biopsies from fish in the summer and calculated the capillary supply (in this case as length density: the length of capillaries per unit volume of muscle) and also an index of oxygen demand (the volume of mitochondria per fibre). As the temperatures fell during autumn, there was a cold-induced angiogenesis, the capillary supply was expanded. As the photoperiod shortened and the temperature dropped further, there was an increase in mitochondrial content (mitochondrial biogenesis). In the process, this increased the girth of muscle fibre and as a consequence reduced the capillary density (capillaries per unit area of muscle); during spring there was an intermediate stage on the way back to summer (Figure 7.6(a)). This cyclical growth pattern coordinates supply and demand, giving a staircase-like advance in aerobic capacity; first grow capillaries to accommodate cold blood, then grow more mitochondria to compensate for reduced activity, then start all over again.

In general, fishes accommodate a large range in environmental temperatures and differing lifestyles by adjusting structural elements that determine the aerobic capacity of their swimming muscles. Indeed, if we look at biopsies from a range of different species and match their oxygen

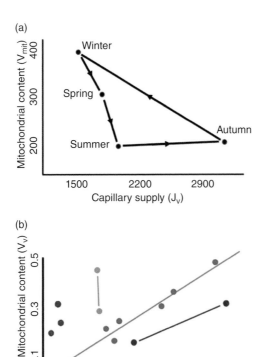

(a)

(b)

FIGURE 7.6 Seasonal and evolutionary differences in capillary supply. (a) Seasonal cycle in growth of capillaries (J_V, capillary length density; mm^{-2}) and mitochondria (V_{mit}, mitochondrial volume; μm^3) in rainbow trout red muscle, reflecting the relationship between oxygen supply and oxygen demand. Note that, during winter, trout appear to be 'under-supplied' with capillaries, whereas they have an apparent excess of capillaries in summer and autumn. After Egginton et al. (2000). (b) Interspecific regression obtained from published data: plotting mitochondrial volume density (V_V) against numerical capillary density (CD) for slow oxidative muscles in a number of fish species demonstrates a general matching of oxygen supply and demand (pale blue). Crucian carp adjust both elements on cold acclimatisation (purple), whereas striped bass appear not to adjust capillarity (green), while other species, including those from the Antarctic, appear to emphasise mitochondrial content (red). Modified from Egginton and Sidell (1989).

supply and demand, we get a linear relationship between capillary density and mitochondrial content (proportion of fibre made up of mitochondria) with only minor variations (Figure 7.6(b)). However, some species show an impressive scope for adapting to environmental changes.

When we cold acclimatise ectotherms, is there a similar response to that shown by endotherms? Recall that, if humans drop their core temperature (T_c) by just a couple of degrees, we enter the realm of clinical hypothermia and require attention, but the Crucian carp can adjust its physiology to cope with an order-of-magnitude greater difference (28 vs. 2 °C) by adjusting both oxygen supply *and* demand. There is a little offset from the interspecific pattern, but capillarity and mitochondrial content are regulated without any apparent difficulty: they remain active and remodel themselves to accommodate cold blood in winter.

Other strategies also exist. We studied the response of striped bass that live off the eastern seaboard of the USA acclimatised to 25 °C (summer maximum) and 5 °C (winter minimum). Cold acclimatisation greatly increased the mitochondrial content, as seen earlier, as well as intracellular lipid depots (a fuel for mitochondria), but there was no apparent change in capillary supply. Examining muscle fibres under the electron microscope, we see, relative to mammalian muscle, lots of capillaries supplying the oxygen fuelling a large population of mitochondria that generate ATP to allow myofibrils in between to contract when they swim. Interestingly, on cold acclimatisation the volume of myofibrils (contractile elements) does not change, so in packing extra mitochondria and extra fuel the fibres have grown bigger. If the number of capillaries is unchanged and the size of fibres increases, then the capillary density (CD, the number per unit area of muscle) would fall – but it does not. This tells us that, although the CD is unchanged, this is not because the microcirculation is unresponsive to cold exposure; quite the opposite, angiogenesis must be matching the growth of fibres, while the increase in mitochondria parallels an increase in capillaries: there is an extremely tight coordination of structural remodelling. A final variation of the basic pattern is that, like cold-acclimatised bass, some species routinely show a higher mitochondrial content than expected from the structural relationship seen in other species. Some of the fishes with a relatively high mitochondrial content but a very low capillary density are those that live in the Southern Ocean around Antarctica. If you want to look at cold-adapted species *par excellence*, this is the place to go.

Life in the Slow Lane

From documentaries about Antarctica, many know it as the highest, driest, windiest, and, indeed, coldest continent. It cooled during continental drift about 22 million years ago when the tip of the Antarctic Peninsula separated from the tip of South America. Once that happened, a circumpolar current was established, which isolated the marine fauna from all other oceans in water that has been constantly cold for at least 10–14 million years. The continental air mass was isolated as well, and its cooling killed off virtually everything there; a mass extinction. A similar thing happened with cooling of the Southern Ocean, there was a massive species crash in which most fish were lost. Because the circumpolar current acts as a polar front, a physical barrier preventing immigration and emigration, what was left was mainly of one type, the notothenioids that now dominate the fish biomass. This group of fishes was able not only to survive, but also to radiate and fill vacant ecological niches. Charles Darwin would have been very familiar with this concept, because this is what he described in the Galápagos Islands, with vicarious migration of finches to the islands and new-found available ecological niches, where they radiated and developed into new species.

Such endemic speciation is what happened around Antarctica; few of these fishes are found anywhere else. There is now a range of species that are genetically very similar, so we can look at the influence of different challenges and different adaptations on the basis of close genetic relatedness. This is important because, when we compare the thermal biology of these fishes with that of temperate-zone species and of those in the Arctic, we see different patterns – but do not know to what extent this is because of the variable or characteristic under investigation, or if it is due to their phylogenetic history (e.g., they may show different responses, be different animals because of their different history). Studying Antarctic fishes removes some of this interspecific variability, and from the geological record and molecular clocks we know this is a case of evolution in progress. Importantly, looking at closely related animals can help focus attention on the physiological responses to environmental challenges, and better understand the origins of ecological phenotypes. In particular, from a thermobiological point of view, sea temperature around Antarctica hovers just above the freezing point of seawater ($-1.86\ ^{\circ}$C) all year round. This is one of the most stable thermal ('stenothermal') environments on the planet, so the extreme seasonality is based on photoperiod rather than differences in temperature. Despite this lack of a normal hibernation cue, fishes anticipate the onset of winter and enter a period of dormancy (Campbell et al. 2008).

Some adaptations to life in the slow lane are both interesting and potentially important. The inshore fish are super-cooled: the temperature of their blood is lower than the freezing point of biological fluids. This happens in the Arctic as well. A famous experiment many years ago put an ice cube in a super-cool aquarium; it started to break up and, as soon as they were touched by a fleck of ice, the fish rapidly froze solid – contact with ice triggers a nucleation event. If not exposed to ice, e.g., in the deeper parts of the Southern Ocean, there is no problem, but life in shallow inshore waters needs some protection from ice crystal growth. Yet we observed species like *Trematomus* taking a rest on some platelet ice at the bottom of the Ross Sea: it can do so only because it generates its own antifreeze. Similarly to putting antifreeze in car radiators to prevent freezing in very cold winters, this animal is preventing itself from freezing when in contact with ice. The discovery of these antifreeze proteins has also been useful from a biotechnology point of view, with applications ranging from tomatoes that are free from frost damage to prolonged transport of organs for transplantation (Eskandari et al. 2020).

One important principle in biology is that, if animals are supremely well adapted to a particular niche or unusual lifestyle, that usually comes at the cost of flexibility – they are at a significant disadvantage if the environment changes. We know this applies to Antarctic fishes. These species have evolved in a stable, cold, and oxygen-rich environment, leading to a number of traits that ensure survival in this extreme environment, such as antifreeze glycoproteins (discussed above). However, other traits have been lost that may restrict the capacity to thrive in a warmer future. Indeed, many studies have shown that Antarctic fishes display reduced thermal tolerance compared with temperate fish species; if placed in a domestic refrigerator, most would die of heat exhaustion! The extent and pace of global warming is obviously of concern. The temperature along the Antarctic Peninsula in particular is rising much faster than that in most parts of the globe, at the kind of magnitude (1 °C over the last 50 years) that in the geological past has been associated with extinction events. The question is are these cold-adapted fishes able to resist global warming, or are we on the brink of a second mass extinction in Antarctic waters?

We have spent a lot of time comparing two of the larger species, *Notothenia coriiceps* (a red-blooded fish expressing myoglobin in skeletal and cardiac muscle) and *Chaenocephalus aceratus* (a white-blooded icefish lacking both haemoglobin and myoglobin; Figure 7.7(a)), very interesting animals for different reasons. It is particularly exciting to look at the icefish, where vicarious mutations provide a natural test of the importance of oxygen transport capacity. When I studied biology at university, the textbooks stated that respiratory pigments are absolutely essential for vertebrate life; we cannot live without them, because we need to facilitate transport of oxygen in blood bound to haemoglobin and facilitate diffusion of oxygen in muscle bound to myoglobin. Around 2.2 million years ago one group of fish, the Channichthyidae, lost the capacity to express globin genes, and so they lack haemoglobin in the blood (there are a few vestigial erythrocytes but none containing the respiratory pigment) and expression of myoglobin in the ventricular myocardium (Figure 7.7(b)). Because these animals live in a very cold environment, and oxygen solubility is inversely related to water temperature (the Southern Ocean contains ~1.6 times more oxygen per unit volume than seawater at 20 °C), they survived in the seas around Antarctica, whereas the same mutation in more temperate zones would be likely to be lethal. Other icefish species lost haemoglobin, so they have white blood, but kept myoglobin, so they have a red heart.

There are a number of features in icefish that are striking, and likely to be adaptations required for life in the absence of respiratory pigments. First, looking into their cavernous mouth you see white gills, unlike any other fish, because they are perfused essentially with plasma. Recall we said that the cardiac mass of *Notothenia*, a closely related cousin (congeneric species) that shares a lot of genetic history with icefish, is large for fish because it has to cope with pumping cold blood. When comparing similarly sized animals (1.5 kg), the icefish heart appears to be enormous (Figure 7.7(b)), which is consistent with the need to pump large volumes

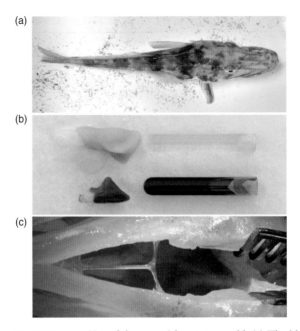

FIGURE 7.7 How fish cope with extreme cold. (a) The blackfin icefish *Chaenocephalus aceratus*; note the large head and reduced musculature of the body. Photo from O'Brien et al. (2021), with permission. (b) The relative ventricular mass of *C. aceratus* (top), a species that lacks both haemoglobin and myoglobin, is three-fold larger than that of a similar-sized red-blooded yellow-bellied rockcod, *Notothenia coriiceps* (bottom). Vials of blood from the two species are shown alongside the hearts; note that the lack of the oxygen-binding protein haemoglobin in icefish gives their blood a milky white appearance. Photo from O'Brien et al. (2021), with permission. (c) A primitive vascular system in icefish (hypobranchial artery) feeds pectoral locomotor muscles ventrally, complementing the normal dorsal aortic supply.

of low-oxygen-content blood around the body, with a disproportionately large atrium. There are also adaptations in the vascular system.

The cardiovascular system in fishes is unlike the mammalian double-pass four-chambered heart, they have a single-pass two-chambered heart: deoxygenated blood comes from the periphery and is pumped through the heart to the gills, then oxygenated blood is pumped around the body, returning to the heart. Notothenioids swim by sculling of their pectoral fins, so the locomotor muscles have a dorsal arterial supply, like all other fishes, but icefish also retain a primitive feature of a hypobranchial system where oxygenated blood comes by another arterial route, reflexed back over the gills: these locomotor muscles have a dual arterial supply, twice the supply of other fishes (Figure 7.7(c)). With blood of low oxygen-carrying capacity (icefish carry around 10 per cent that of red-blooded fish), to maintain the same oxygen consumption, there is a requirement for a great increase in cardiac output. This explains why there is such a large ventricle, but in order to achieve that output efficiently, the resistance against which the heart has to pump, the cardiac afterload, needs to be reduced as well. So we examined vascular casts of the capillary bed in pectoral (swimming) muscles. The intravital microscope images of rat muscle showed capillaries running parallel to muscle fibres (Figure 7.5 (a)) but here they run in an arcade or loop kind of network. More importantly, their size is some four times that of other fishes. As capillary diameters tend to be fairly constant amongst vertebrates, this is an unusual family with very large capillaries, an up-scaling that occurs in other vessels such as venules and arterioles as well, minimising the resistance to perfusion by cold blood.

To examine cardiac function, we used electrical activity of the heart, in a similar way to how doctors use an electrocardiogram trace; I think we were the first to do this with icefish (Figure 7.8(a)). The electrical activity of the heart has a familiar form, being dominated by large R waves (denoting ventricular contraction) because of the very large ventricle, but P waves (atrial contraction) and T waves (ventricular relaxation) are also present. To determine the resting heart rate in a fish is very difficult, because if a door opens in an adjacent building these sensitive animals respond and become tachycardic (elevated heart rate), so recordings were done at the crack of dawn. The first thing of note is the very slow rate – five beats every 25 seconds; this really is life in the slow lane. Just as coordinating a regular cardiac rhythm is hard at very high heart rates, organising rythmicity at very low rates is also problematic. However, there is wonderfully coordinated contraction of atria and ventricles, pushing blood through the bulbous arteriosus to the gills.

Our aim was to determine the thermal sensitivity of these fish that are supremely well adapted to a particular niche. Antarctic fish are very well adapted to stable and very cold temperatures, but what may happen if that ecological niche changes? One of the most common ways of addressing this question is to determine a critical thermal maximum (CT_{max}),

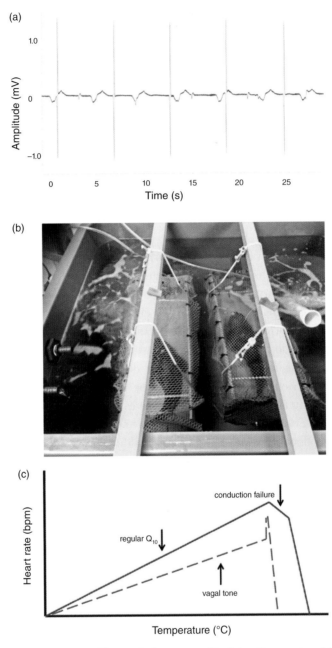

FIGURE 7.8 Changes in heart rate of icefish with warming. (a) The first ECG trace from *C. aceratus*, showing a resting heart rate of around 10 beats per minute at −1 °C (large spikes are R waves, denoting ventricular systole). (b) *N. coriiceps* in

where the water temperature is gradually raised throughout the day until an animal loses the capacity to keep upright in the water (Figure 7.8(b)), giving an integrated response of many physiological mechanisms. The expectation of most people was that icefish would be very sensitive to temperature change, the impression being that they are at the edge of survivability and any additional stress would lead to failure of functional capacity. But while the CT_{max} is indeed a little lower ($>14\,^\circ C$) than that of the red-blooded *Notothenia* ($<17\,^\circ C$), that is still an impressive excursion for an animal that has only experienced $-2\,^\circ C$ water for a couple of million years. If we pharmacologically block autonomic regulation (vagal tone), then any difference between these two species is lost, reflecting their common ancestry (Egginton et al. 2019).

To follow the response of the heart per se, we can look at the breakpoint temperature, the temperature at which the normal rhythmicity of the heart breaks down (Figure 7.8(c)). Using the gradual thermal ramp, we see heart rate in the red-blooded *Notothenia* rise in a fairly linear manner (red line). This is a typical biological rate function, a thermal sensitivity (estimated Q_{10} around 2) for every 10 $^\circ C$ rise showing a doubling of heart rate. This continues until the break point (14 $^\circ C$), at which in other vertebrates we normally see cardiac fibrillation, whereas these animals display a very strange phenomenon (episodic ventricular tachycardia): short bursts of extremely fast beats followed by a period of no beats, gradually these periods of asystole continue until failure of the electrical conduction system, at which point the heart would stop beating. As with hamsters (discussed above), returning the animal to its normal environmental temperature will allow recovery. Icefish are also sensitive to warming; no animal can avoid the pervasive effect of heat causing a rise in heart rate, but the rate of increase is less (blue line) than in *Notothenia*.

CAPTION FOR FIGURE 7.8 (*cont.*) suspended slings to allow unobtrusive recording of ECG and administration of drugs (normally behind a black screen); the tanks have a double-pump system to deliver a controlled temperature ramp. (c) Changes in heart rate of *N. coriiceps* (red) and *C. aceratus* (blue) with warming. 'Regular Q_{10}' refers to the temperature coefficient, which is a measure of the thermal sensitivity of biological processes (usually defined as the ratio of rates at temperatures differing by 10 $^\circ C$). Note the similar breakpoint temperature, but different pattern of response.

These animals try to minimise any increase in heart rate by raising vagal tone, because as temperature increases the oxygen content in water is reduced and the work of extracting oxygen from the environment goes up; a double whammy. They continue with this high vagal tone until just before the break point (13 °C), at which they can no longer sustain the effort, so there is quick onset of tachycardia and heart failure. Again, pharmacological blockade removes even this small difference (Joyce et al. 2018a). Despite arriving by a different route, these species tolerate a fairly similar temperature, suggesting a degree of thermal resilience in these animals that had not previously been recognised.

The final question was whether it is possible to thermally acclimatise these animals. In the case of the icefish that proved very difficult, but the red-blooded species coped with warm water over a number of weeks. After 5 °C acclimatisation there was no change in the temperature at which the first signs of arrhythmia occur in *Notothenia*, suggesting limited capacity for adaptation in electrical coordination of cardiac performance. When *Notothenia* are acutely warmed, both oxygen consumption and cardiac output rose to a similar extent to that in animals at 0 °C, with a similar heart rate response, until ~12 °C. However, warm-acclimatised fish were able to increase oxygen consumption further and tolerate another 2–3 °C rise, aided by increased stroke volume (Joyce et al. 2018b). It appears there is some capacity for warm acclimatisation in cold-adapted fishes, and hence resilience to higher temperatures in the red-blooded species.

We must add the caveat that, because of the challenges of working in remote regions, it is possible to conduct only relatively short-term acclimatisation experiments. So, although this suggests there are molecular mechanisms that may provide these animals with the potential for adaptation to higher temperatures, we cannot test what will happen over a longer time period (such experiments are very difficult to do in the logistically challenging environment of the Antarctic). Neither do we know what will happen with other systems in the body, for example, if the locomotor system cannot adapt as well as the cardiovascular system, these animals may not be able to feed adequately; if the reproductive system is compromised, there may still be a species crash. In terms of global warming, the consensus seems to be that the changes are probably too fast for many endotherms as well as ectotherms to adapt successfully if they are already living close to their thermal limits; there is a real danger of ecological collapse. But the situation appears a little less bleak than before our most recent expedition, offering some hope that the remarkable marine life of the Southern Ocean may be able to adapt, at least in part, to the rapid changes it is currently experiencing as a result of global climate change (O'Brien et al. 2021).

Summary

- For most endotherms, thermal balance is lost outside an optimal temperature range, the thermoneutral zone, where a rapid homeostatic response is seen.
- There is some evidence that humans can habituate to low environmental temperatures, allowing adequate perfusion of cold blood to the periphery.
- Some endotherms are able to go one step further and reset their hypothalamic thermostat, and regulate low core temperatures.
- This involves compromise in the cardiovascular system: cardiac and vascular remodelling, and differential adaptations in tissues according to aerobic capacity.
- Ectotherms share a number of common features on acute cold exposure, including bradycardia and cardiac hypertrophy.
- Preservation of oxygen transport to tissue involves multi-level adaptations, including changes to the microcirculation.
- At the thermal extreme, Antarctic fishes show some remarkable adaptations to living with cold blood.
- Even in a species with relatively low thermal tolerance, such as the icefish, some physiological adjustments may be possible to withstand elevated temperatures.

Acknowledgements

It is not possible to do this kind of work without the collaboration of fantastic colleagues and students, and I gratefully acknowledge their input and assistance. Of course, wilderness science is very expensive, so without the generous support from various funding bodies none of this would have been possible.

References

Bell, R. M., Mocanu, M. M., and Yellon, D. M. (2011) 'Retrograde heart perfusion: the Langendorff technique of isolated heart perfusion'. *Journal of Molecular and Cellular Cardiology* 50, 940–950.

Campbell, H. A., Fraser, K. P. P., Bishop, C., Peck, L. S., and Egginton, S. (2008) 'Hibernation in an Antarctic fish: on ice for winter'. *PLoS One* 5(3), article e1743.

Egginton, S., and Sidell, B. D. (1989) 'Thermal acclimation induces adaptive changes in subcellular structure of fish skeletal muscle'. *American Journal of Physiology* 256, R1–R9.

Egginton, S., Cordiner, S., and Skilbeck, C. (2000) 'Thermal compensation of peripheral oxygen transport in skeletal muscle of seasonally acclimatized trout'. *American Journal of Physiology* 279, 375–388.

Egginton, S., Fairney, J., and Bratcher, J. (2001) 'Differential effects of cold exposure on muscle fibre composition and capillary supply in hibernator and non-hibernator rodents'. *Experimental Physiology* 8, 629–639.

Egginton, S., May, S., Deveci, D., and Hauton, D. (2013) 'Is cold acclimation of benefit to hibernating rodents?' *Journal of Experimental Biology* 216, 2140–2149.

Egginton, S., Axelsson, M., Crockett, E. L., O'Brien, K. M., and Farrell, A. P. (2019) 'Maximum cardiac performance of Antarctic fishes that lack haemoglobin and myoglobin: exploring the effect of warming on nature's natural knockouts'. *Conservation Physiology* 7(1), article coz049.

Ellis, C. G., Milkovich, S., and Goldman, D. (2012) 'What is the efficiency of ATP signaling from erythrocytes to regulate distribution of O_2 supply within the microvasculature?' *Microcirculation* 19, 440–450.

Eskandari, A., Leow, T. C., Rahman, M. B. A., and Oslan, S. N. (2020) 'Antifreeze proteins and their practical utilization in industry, medicine, and agriculture'. *Biomolecules* 10, 1649–1667.

Hammel, H. T., Elsner, R. W., LeMessurier, D. H., Anderson, H. T., and Milan, F. A. (1959) 'Thermal and metabolic responses of the Australian Aborigine exposed to moderate cold in summer'. *Journal of Applied Physiology* 14, 605–615.

Hauton, D., May, S., Sabharwal, R., Deveci, D., and Egginton, S. (2011) 'Cold-impaired cardiac performance in rats is only partially overcome by cold-acclimation'. *Journal of Experimental Biology* 214, 3021–3031.

Joyce, W., Egginton, S., Farrell, A. P., Crockett, E. L., O'Brien, K. et al. (2018a) 'Exploring nature's natural knockouts: in vivo cardiorespiratory performance of Antarctic fishes during acute warming'. *Journal of Experimental Biology* 221(15), article 183160.

Joyce, W., Axelsson, M., Egginton, S., Farrell, A. P., Crockett, E. L. et al. (2018b) 'The effects of thermal acclimation on cardio-respiratory performance in an Antarctic fish (*Notothenia coriiceps*)'. *Conservation Physiology* 6(1), article coy069.

Kilgore, D. L. Jr, and Schmidt-Nielsen, K. (1975) 'Heat loss from ducks' feet immersed in cold water.' *The Condor* 77(4), 475–517.

Nash, G. B., and Egginton, S. (1993) 'Comparative rheology of human and trout red blood cells'. *Journal of Experimental Biology* 174, 109–122.

O'Brien, K. M., Joyce, W., Crockett, E. L., Axelsson, M., Egginton, S. et al. (2021) 'Resilience of cardiac performance in Antarctic notothenioid fishes in a warming climate'. *Journal of Experimental Biology* 224, article 220129.

Reinertsen, R. E. (1982) 'Radio telemetry measurements of deep body temperature of small birds'. *Ornis Scandinavica (Scandinavian Journal of Ornithology)* 13, 11–16.

Wunderlich, C. A. (1868) *Das Verhalten der Eigenwärme in Krankenheiten.* Leipzig: Otto Wigand.

Young, S., and Egginton, S. (2011) 'Temperature acclimation of gross cardiovascular morphology in common carp (*Cyprinus carpio*)'. *Journal of Thermal Biology* 36, 475–477.

8 Blood Sculptures

MARC QUINN IN CONVERSATION WITH IOSIFINA
FOSKOLOU

This interview has been edited and shortened for clarity.

MQ   Hi, I'm Marc Quinn, I'm an artist, and I live and work in London.

IF   First of all, thank you very much for agreeing to give this Q&A for us. Let's talk about your background and why you became an artist.

MQ   I don't know if you become an artist. I guess I always was an artist. I was always interested in making things as a way of understanding the world and interested in looking at art. I think it's more of a realisation that it's possible to be an artist and that you can do it as a full-time occupation, so that development gradually happened over the period I was at school and then after I left school.

IF   Was it important for you to study art history in Cambridge in order to become an artist?

MQ   Yeah. I mean I kind of decided not to go to art school. I was a bit headstrong. I was at school and found myself doing Cambridge exams and I thought, actually in art school I'm not really interested in someone telling me how to make art, which is what I thought would happen at art school. Essentially someone's telling you how to make art, but it would be interesting to know about what's happened before and to be able to find out for myself for context for my work. I'm not obsessively interested in art, so just watching, looking at, reading books, and going to lectures about the history of art seemed like a much better idea and I could do stuff on my own.

IF   Interesting. So, can you tell us a little bit about your practice as an artist?

MQ   I just come to work at the studio and make things and do shows.

IF   That's great, that sounds a very good way to do what you love.

MQ   But yeah, I've got different things. I make lots of work. I'm making paintings about the news now which are going to be in a show next year. Next year is very busy because everything from this year has been postponed to there and then I've got a philanthropic side to my work as well. I have been working with refugees for the last five years and

170

making art with refugees, which I'm sure we'll talk about later, so lots of different things.

IF   Yes indeed. As you said your work is connected to blood and I was wondering, as I am a biologist, and I was always fascinated about blood but for different reasons: what fascinates you about blood?

MQ   When it came to making the first work of blood itself, I think what happened was this idea that art was too divorced from reality and divorced from life and it was becoming a kind of game. It's like people making more and more esoteric things for fewer and fewer people to look at. I felt like I wanted to bring some reality back into the world and for art to be about the lived experience of being a person in the world. Then I was also interested in portraiture and I studied Rembrandt at Cambridge and always loved classical antiquity and classical sculpture. It struck me that you look at the reality of a Roman portrait head and I thought: what's the next step, how can you get more real than that? And my idea was if you could somehow make a sculpture out of the actual sitter that would be really interesting. Obviously, I don't want to kill people and then I was thinking about all different parts of the body and different organs. I was thinking about blood, and I thought that actually I love its paradox: it's this material that really keeps us alive, that brings life all to all parts of our body and that remakes itself, which is amazing.

Not just symbolically but literally: what is this life? Blood is usually associated with death and violence and acts and killing and illness, so I thought it was quite interesting. As a material for art, it's great because my body remakes it. I realised quite early on that I would have to be the subject at the beginning anyway because I couldn't ask anyone else to do it. I also felt that if you're an artist wanting to explore what it is to be a person in the world you've got to start with yourself. Blood is really interesting. It's this product that my body makes, and I can give blood and not die – that's strange – but then how do you make a sculpture out of it, because how do you turn the liquid into a solid? So, there's the physical challenge. I'd also constantly been thinking about freezing things as a way of making sculpture so the two came together in a moment of having been thinking about all these different things separately. One morning I woke up and just had the idea: well, why don't I try and make a frozen head out of blood? Yeah, that sounds like a really good thing.

IF   How easy was it; what technical difficulties did you encounter?

MQ   Enormous. The first thing was how to get the blood out of my body. I was kind of a broke artist. I didn't know anyone, but a friend of mine who is a sort of collector – I talked to him, and he knew this doctor who might be able to help. So, I went to see this doctor and I pitched him the idea. I said: I'd like to make the sculpture of my head and I've measured the volume in my head and it's ten pints, so I need to come and give ten pints of blood. Would you do that? He thought about it for a while and he agreed to do it. This time had, you know, slightly less draconian health and safety rules than there are now. So, I started to give the blood.

At the same time, I had 10 months to work out how to make the sculpture. I worked with a guy who made bespoke freezers mainly for morgues and the food industry, and I started to make some prototypes with him. I obviously had no budget to do any of this so it was really on a shoestring, and he made this freezer where you could keep the object displayed at $-18\,^{\circ}$C. We put Perspex hoods around it, it was a big rush, and then on the day of the exhibition I defrosted all the blood I'd given in the last year and poured it into the mould. It had a base plate on it, and then I took them all off and put it in the machine. There it was, the frozen head, and it seemed fantastic to me. Then Charles Saatchi bought it on the first day and everything was great. After about a week, the guy who ran the gallery said 'I think you should come in have a look'. I came in and there was a kind of dust under the chin, and I was thinking 'What the hell is this?', so I called my dad, who is a scientist. He's actually a Fellow of Jesus College and the Royal Society, so I asked him and all his friends and we realised that the freezer was freeze-drying the head. Basically, all the moisture was being sucked out because in a frozen system moisture always goes to the coldest bit, which is the evaporator of the machine. Basically the head was freeze-drying and eventually would just become a little mound of powder, which for another piece would be interesting but not for this one.

So, what happened was this piece was on show for a month or however long it was, and I couldn't work out this technology in that first month, so every weekend I had to go into the gallery at four o'clock in the afternoon on Saturday, or whenever it closed, and melt down the head, recast it and put it back in. I think they closed on Monday. So, melt down the head, recast it, refreeze it, put it back in the thing. It would just about last for five days until the next weekend. I was also an alcoholic then so it was a bit chaotic, but I managed to do it. And then, after this kind of nightmare period, I then had a year and a half or something to work out what to do before Charles Saatchi showed it in his museum. It went through various iterations until finally we settled on the head now, which is in a box of liquid silicone oil that serves as a barrier to stop the air getting to the blood and being able to evaporate it. The oil doesn't attack the blood, so that was the final solution. Because liquid silicone oil stays liquid to a much lower temperature than $-18\ ^{\circ}$C, where the head set, that solved it, but it was quite a journey.

IF   So, does this temporal nature of the sculpture excite you? For example, if the electricity went off the sculpture would completely melt.

MQ   Yeah, I think it's very much about dependence. You know, probably unconsciously my dependence on drinking came into it as well. The idea that made me realise this idea of vulnerability. We're on zoom, I've got the lights on, people are all here because they came around or whatever. It's this whole dependence on infrastructure and on each other that is implicit in the work, which I really liked as well, so even though it's called *Self,* it's really about how we don't exist without other people. It's about this idea that everyone is connected, and I think we've seen that more than ever this year and that we can only really look after ourselves by looking after other people. I think that's what led to the subsequent blood sculpture that we'll talk about later.

IF   I'm sure, yes. So first, what do you think that blood says about society? And then I would really like to talk about your new project, *Our Blood.*

MQ   It's about how vulnerable every human life is, and I like the idea that it's a bit like life. You can say someone's alive, and then they're dead, and where have they gone? You have no answer. You can just say they were there and now they're not. In a way, the form of the sculpture when it's frozen is an emerging property like life, and then when you turn off it becomes a pool. Where's the form of this head? It's just not there anymore. You can't say it's gone somewhere: it just doesn't exist. So, I had a nice metaphor for life and death but also it's the whole series. It is a series that was conceived from the beginning, that I would make one every five years. I think that was again sort of inspired by Rembrandt and his self-portraits that he did during his life. I did a show in 2009 with the Beyeler Foundation in Switzerland, where there were four sculptures and me in the room. There was 50 pints of my blood in the room, and to me that really underlies how it's about the way that life regenerates itself, the idea of regeneration, new beginnings, and all this. So, I think there's an optimistic side to it even though it may appear to be about violence and death. I like things in the paradox club that appear to be one thing and then on close examination turn out to be something almost entirely opposite. So, I think that it doesn't have an answer, it doesn't have one meaning, which is probably why it's still interesting for me to make them because every time I do it, I learn something new and its meaning develops and changes over time.

IF   This year is 30 years since the first blood sculpture, right?

MQ   Yeah, I think this year is 30, so I think I'm going to do a show next year, because we're continually upgrading the technology. I've got the blood, but I didn't make the sculpture because we've been working on this three- or four-year project to remake the freezer device in a better way, and we're just about to finish that. So, the two new ones will come out and we'll do a show somewhere of all seven together.

IF   And what about the new project you're working on, *Our Blood*? Can you tell us a little bit about it?

MQ   I think *Our Blood* started about six years ago. It was about seeing people, refugees, being forced to leave their homes from war or famine or any number of other reasons why people leave and being sort of dehumanised and just being seen as masses of people either squashed on boats or walking across Europe. One of the problems was this idea that people have become dehumanised by the process of becoming refugees. Having met many refugees during the project (everyone from doctors, schoolteachers, to anyone you'd care to meet here, you'll also meet in the group of refugees), I thought it'd be wonderful to make something with contributions from thousands of people: to make something about humanity coming together and sticking together and supporting each other.

I thought it would be really interesting to make a cube: something completely geometric and abstract that's really quite big, like a metre cube. If a few thousand people gave a really small amount of blood, you'd have enough to build this thing. It's like a library, you know; it has all the blood and all the DNA, it has all the ancestry, it has all that. You could take that cube and take it to another world, and they would know everything about us from there. It would be like the Rosetta Stone of humanity or something, so it felt really interesting to make this kind of index or library of human potential. Originally, I was going to mix everyone into one block, and then I thought it was actually stronger to have the idea of people standing together, a bit like the idea of the cell, so I decided to make two blocks, and one was given by refugees, and the other was given by people who don't see themselves as refugees, but neither is labelled. You can't tell the difference between them by looking, so it's this idea of difference and it doesn't exist visually. It doesn't really exist either, in that it'd be very difficult to tell which was which even if you were doing genomes with other people because everyone's everywhere anyway. So, it's the idea of solidarity, standing together, diversity, and strength. As time's gone on those things have just become more and more interesting and relevant.

It's a long project because we had to form a non-profit charity and raise money to make the object. Any money raised by the whole project goes directly to refugee charities, and we're working with the

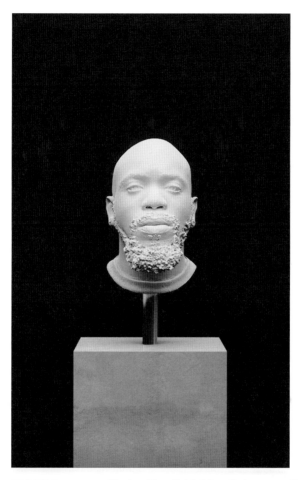

FIGURE 8.1 *100 Heads – Yves Saidi*, Marc Quinn 2021. Copyright and courtesy Marc Quinn studio.

International Rescue Committee,[1] the world's biggest refugee charity, which will get half of the money we create. The other half will go to smaller refugee charities. So, we had this huge project. In the meantime, it's quite interesting because I took myself and split it in two, so you had the blood bit, those cubes in *Our Blood*, and then I started to make portrait heads of refugees in concrete (Figure 8.1) to benefit the project

---

[1] www.rescue-uk.org.

as well. It's almost like I took it and separated the portrait head bit and the blood bit, but it's all one piece, so it's very connected to itself but it's really about everyone. It seems like such an unarguable idea, you know, that my blood and your blood and everyone else's blood is essentially the same. Obviously, there are individual differences, but it's the same stuff.

I believe that art can communicate in a way that that sentence can't. In a sense people go 'yeah yeah yeah', but when you see a cube of it, a metre square right in front of you, and maybe you put your phone on the QR code and see and have the life stories of every single person in that cube, and listen to them, watch them, then it's a different level of really understanding. A different visceral level of connecting than just the idea, I think. That's why I wanted it to be a public artwork as well, so it's going to be on the street in a kind of pavilion that Norman Foster's designed, and anyone can go and look at it. You don't have to pay to go in. You can make a contribution to the charity if you want, but it's going to be free. Just something landing in the middle of daily life, people walking down the street, wondering what it is, and then going in to see it. That way of connecting with people is incredibly powerful. I'm using the public realm to talk about this idea, which is essentially about people and how we deal with countries. You know, how to make us think a bit more humanely. It's very much a humanitarian thing rather than a political thing. It's one of those things where you just do it and see what happens, and then in our times it's really good to have something to bring us together, when in a lot of countries we see the exact opposite and a lot of extremes.

IF You have already answered a lot of my questions. How was it to go and ask refugees to give you their blood: how did you do it?

MQ When I first had the idea, I thought it was probably not an idea that could ever happen, because when people are refugees the last thing they want to do is give you some blood. I immediately realised that I wouldn't be taking blood from people in refugee camps. I'd be taking blood with people who have settled. Refugees in countries when they've arrived, so I went to Berlin, where I had a contact, and I set up a meeting with representatives of many different refugee communities. I pitched the idea, and if they had said it was completely ridiculous, I would have just not done it. The idea was that we can give this small amount which then our body remakes and it empowers us, and it connects us, and then we become part of this idea about our own destiny. The other thing the refugees I spoke to said is that people are very, very squeamish on their behalf, but actually they're totally into it, so 'yes'. After that I felt like it could work, and that was subsequently borne out by all the other refugees that have been amazing collaborators and advocates that we've met and worked with over the years.

The pandemic has obviously stopped being able to do quite a lot of work on it. I think [the pandemic] also made it a more interesting project in a way. Unfortunately, these ideas and these problems are not going away. The other thing I always felt is that I came to an idea that was moved by actually seeing the people, the refugees. In around 2015, I realised I had given to charities to do with climate change, and somehow this topic was about all those things too, because those are also some of the reasons that people have to leave their homes. It's amazing, so you can talk about all those topics through the lens of refugees and therefore a human lens. I think what's interesting is keeping it about people, because the only thing that cares about what happens in the world is people. The world itself, you know, geography, plants, just keeps on going, and so people and connections between them are really the only thing there is, and I thought I wanted to keep it really human.

IF   I wanted to ask you why you chose the refugee crisis and not the climate crisis to work on, but you just replied to that, so what about social justice? Your work is quite political and it seems that doing public work is very important to you.

MQ   I think my art isn't political, it's more about people and social justice, or just humanitarian issues. I think that doing work in the public realm has a different resonance to doing work for art galleries, and I think that as a successful artist it's kind of my duty to use that privilege and power to make a difference, to start debates and start conversations, and to do things. I think that I have higher ambitions for art, that it can change the world, that it can be involved in the great debates of our time in a very unique way because it's not really verbal and it connects. You don't have to understand the language. You don't have to translate it. People can understand it from any background or place, just being there with themselves and their experience.

When I did the sculpture of Alison Lapper in Trafalgar Square,[2] there was almost a situation where people said I shouldn't do it because the public wouldn't understand. I realised this was an outrageous prejudice. As soon as it started, we had a whole lot of different views from lots of people, but all those views were articulated properly. They're mostly all positive, but even the ones that were negative showed that it made people think for themselves, and they should think for themselves. It's interesting to make the public part of debate and to slightly bypass the

[2] *Alison Lapper Pregnant* (2005).

whole media thing where people have been told what to think. Let people make their own mind up.

IF   So what about the Paralympic Games? How did you think of this great sculpture?[3]

MQ   The sculpture of Alison Lapper again came from a long series. A lot of it has long gestation. I started in about '98 or '99. I'd been in the British Museum looking at the fragmented marble statues of the Elgin marbles and thought about how people say these are the most beautiful sculptures. Then I was in Paris and all the tourists were photographing themselves in front of the Mona Lisa, so I'm in by the Venus de Milo and I thought it was really interesting that this is accepted and in fact stands in for nostalgia. You know, it's almost like people empathise because of the lost pieces, but if a person who is born or by accident or illness loses their limbs or parts of the body, you were, particularly at that time, going to get a very different reaction. It just seemed like 'Wow, that's such a massive gap'. I got a very simple idea to find someone, for instance a Paralympian, to follow the tradition of ancient sculpture wherein marble statues of athletic heroes were made, but to change the idea slightly by working with a disabled athlete, and so I started by making a sculpture of Peter Hull. Actually, I've got a copy of it here [showing a replica of the sculpture in the studio].

So, Peter Hull won a gold medal in the Seoul Olympics in swimming and he's an amazing guy. He works for Hampshire County Council on sporting facilities for disabled people. He's totally, completely self-sufficient. He's also someone who actually said yes when I asked him 'Would you sit for a sculpture?' I made that sculpture in Macedonian white marble. It felt like a sculpture from a different society where differences were celebrated rather than being hidden.

Then I decided to take this idea further and I made 10 sculptures, portraits of different people, and these were shown in so many places including the Victoria and Albert Museum. After that there was the competition for the fourth plinth, and I met Alison, who is Peter's best friend. When I started the series Alison had a bad back and she couldn't make the sculpture, so we didn't work together, but then about a year and a half later she called me up and said she was pregnant so maybe it was a really good time to do a sculpture. I was like, yeah, absolutely let's do that, so together we made the sculpture of her, a life-sized marble of her pregnant. Then, when the fourth-plinth competition came along, I thought it was a perfect antidote to the square, which is full of dead white blokes who conquered the world. I thought, why not have a

---

[3] *Breath* (2012).

woman who's overcome so much and is an amazing artist and mother. Also, all those sculptures are about the past, and a pregnant person is always about the future, so it's a monument to the future in a way, and monument to Alison's amazing strength. Celebrating difference in how everyone can be beautiful. And so I made the proposal and amazingly that got through and then that went on the plinth.

Subsequently, when the Paralympics were in London, they asked me to make a large version of it for the opening ceremony, which was amazing. The first one in Trafalgar Square was made of marble. I think it's interesting that the Paralympics one is four times as big. [They said] we're going to have Stephen Hawking at one end and the lights will go across Stephen Hawking. They come back and the sculpture will be there. I thought, this would have to be a huge marble sculpture, you can't just bring it on in three seconds. They said it's fine, we're just going to go from Stephen Hawking to your sculpture, so that's okay. Then I needed to rethink a little bit. What usually happens is some kind of restriction can make a new thing happen, and so I looked at different ways of what I could do. I could have made a big, lighter glass one or something, and I looked at inflatable things, and then I found this company. This was 10 years ago, and there was almost no scanning. This stuff was around but it wasn't that prevalent. I found this company that could scan the original sculpture into the computer and make a fabric copy that had the photographic imprint of the sculpture on it as well. When it's blown up, it looks like the sculpture, but different, and then, when you're close, it has all these seams on it. I mean, you can touch it, it's like a living thing, you know: it's soft and warm because of all the air moving, so it felt like it's always like a colossal sculptural version of *Self*. It's this big form, this huge vulnerable and yet enormously powerful figure, but if you unplug it, it disappears, so it's just like the blood disappears. Suddenly this seems to be not only a great technical way to do it, but it also made a really good different sculpture called *Breath*. As well as being in the Paralympics it was shown in the Venice Biennale in 2013, in front of the Church of San Giorgio.

IF Very nice. I really liked what you said about having a very different sculpture in Trafalgar Square, where you depicted a disabled pregnant woman, surrounded by old sculptures of usually privileged white men. It reminded me of a current work you had with Black Lives Matter. Do you want to talk a little bit about this?

MQ That piece was in July 2020.[4] I've been following the news closely through this period as I'm making all these paintings about the news.

---

[4] *A Surge of Power (Jen Reid)* (2020).

And of course, the Black Lives Matter movement was a huge part of the news, following the horrific killing of George Floyd. It became essential for everyone. It just became bigger and bigger and bigger, and then, when the statue of Edward Colston was taken down, I thought it was really interesting that it created this amazing, charged place on the top of that plinth. I wasn't thinking of making it, and then a couple of days later I saw this picture from the Internet of Jen Reid standing on the plinth with her arm up in a black power salute, and it struck me that she created the sculpture that should not remain permanently but for a moment should go and reactivate that space. So I sent her a message on Instagram and asked if she would be interested in collaborating to make a sculpture, and she was very enthusiastic.

She's an amazing person, very strong and definitely knows exactly what she thinks about everything. I thought maybe she wouldn't want to be involved at all, but she was incredibly enthusiastic and a great collaborator, and so, very, very quickly, over a month I made the sculpture with her recreating the pose. She wore the same clothes that she'd worn, she did everything, and I made a resin one with a very heavy base. I got various people to look at how one could actually put the sculpture on there safely, and do it with health and safety, and do it all very professionally. The only difference is that we didn't tell anyone we were doing it. At one point I thought that maybe we should collaborate with Bristol City about it, but then I realised it was an impossible situation for them. So Jen and I and my team were down there, and at five o'clock in the morning the sculpture was put on the plinth by my team.

That was completely fine, nothing happened, they left, and then the whole world sort of blew up and it created this enormous storm of debate from all sides, and then 24 hours later it was gone. But during that 24 hours, people were even there at 8 a.m. that first day, people were bringing their kids on the way to school to look at it, and there was a kind of festival with people like playing music and stuff around it. It was incredible how it activated everything. I think 24 hours in the age of the Internet is almost like a year in Trafalgar Square pre-social media. So, it went around the world virtually, and now I think that it's going to end up in a museum, but we'll know that next year. That's what seems to be the direction, and so I think that that would be a great place for it to be, because I don't think it should be shown in a public space again. That was really very site-specific, and it was that moment in that place, and now it should be indoors somewhere, like a relic of a performance. It's almost like a performance. It was never intended to replace or be there forever. It was about looking at charged places and publicly activating them, even if just for 24 hours.

Subsequently, Jen's had a whole new career, talking and going to schools, and she's been interviewed in the media, and sort of amazing things have happened to her life since it happened, which I'm so happy about. We're still obviously great friends and talk a lot.

IF   It seems to me that depicting history as it happens is also important to you. Do you think that as an artist you have an obligation to record current affairs to educate the next generation or to keep a record?

MQ   I don't think I have an obligation. I think it's what really excites me . . . I also studied history at university. I really like history and I really like art, and I think there's this really charged moment where the two come together, and I think that history painting in the early eighteenth or nineteenth century is the highest form of art, but was obviously very much top down. Napoleon crossing, or an emperor on a battlefield with everyone supplicant below him . . . and then in around 2011 I started to look at the protests after the killing of Mark Duggan in London where there were these extraordinary pictures of protesters, and I thought this is history painting but just someone's turned it upside down. It's history made by people coming up and getting on the streets and doing stuff about it. So, I started painting series from then, and that's continued, and then this year when the lockdown happened, I was here in my studio on my own and I thought about what I could do. I was looking at the news and looking at my phone really, and doing screenshots of things that interested me, and then I suddenly realised that I couldn't spend six months painting a painting like I had done before. There was so much I could do, something every day, and that those pictures should be based on screenshots, so they're kind of like paintings of screenshots. It's like a mad rush through everything that's happened in the last year.

IF   So it is very Covid-related.

MQ   Not Covid only. I mean it includes everything that's happened – it's really about everything that happened last year from Britney Spears to Tokyo. Everything that is important has had a relevance, and that I found interesting. Anyway, it's like my diary; like a diary of the year.

IF   Sounds very interesting. Since we're in the Darwin Lecture series, what is evolution for you and how has your work evolved through the years?

MQ   I think that that's how art works. I think from looking at what we've said, pieces evolve from other pieces, and ideas evolve from other ideas, and also evolve in a sense which is quite Darwinian perhaps, evolving from constrictions and from environment. I mean, the environment shapes something, but then it's the same thing when you can't do something, you have to think of another way to do it, and so often a constriction or restriction can evolve a new way of doing something, and a new way of seeing things. But you know, it's all about being a human as all of Darwin is as well.

IF And maybe in the future we can also bring animals into your projects.

MQ Who knows?

IF Who knows. Great. Well, thank you very much, that was absolutely great. I really enjoyed it.

MQ Great, thank you so much as well.

Index